A Guide to Practical Stereology

Karger Continuing Education Series

Hans Elias, San Francisco, Calif.
Dallas M. Hyde, Davis, Calif.
with a contribution by
Richard L. Scheaffer

A Guide to Practical Stereology

148 figures and 6 tables, 1983

Basel · München · Paris · London · New York · Tokyo · Sydney

Hans Elias, Ph.D.
Professor Emeritus, Chicago Medical School; Former Guest Professor, University of Heidelberg; Research Associate, Department of Pediatrics and Pathology, University of California, San Francisco; Founder, First President, now Honorary President, International Society for Stereology.

Dallas M. Hyde, Ph.D.
Assistant Professor, Department of Anatomy, School of Veterinary Medicine, University of California at Davis, Davis, California.

Richard L. Scheaffer, Ph.D.
Chairman, Department of Statistics, Nuclear Sciences Center, University of Florida, Gainesville, Florida.

Karger Continuing Education Series, Vol. 1

Topics covered in the Karger Continuing Education Series are selected to help improve clinical skills and introduce the reader to health-related areas undergoing exceptional growth. Produced as compact instructive texts, volumes set forth information which serves to heighten the general awareness and command of current medical procedures and practice. The concise textbook format enhances the value of these books as convenient teaching and training tools for medical scientists, medical clinicians, and health professionals.

National Library of Medicine, Cataloging in Publication
   Elias, Hans Michael, 1907-
   A guide to practical stereology / Hans Elias, Dallas M. Hyde, with a contribution by Richard L. Scheaffer.—Basel; New York: Karger, 1983 (Karger continuing education series, v. 1)
   1. Biometry   2. Histological Technics
   I. Hyde, Dallas M.   II. Title   III. Title: Stereology
   QS 525 E42g
   ISBN 3-8055-3466-3

All rights reserved.
No part of this publication may be translated into other languages, reproduced or utilized in any form or by any means, electronic or mechanical, including photocopying, recording, microcopying, or by any information storage and retrieval system, without permission in writing from the publisher.
© Copyright 1983 by S. Karger AG, P.O. Box, CH-4009 Basel (Switzerland)
Printed in the United States of America
ISBN 3-8055-3466-3

Contents

Dedication vi
Preface vii
Acknowledgments ix

     I. Some Elementary Geometric Concepts.............. 1
    II. Basic Definitions, Terminology, and Symbolism...... 16
   III. Stereological Measurements of Isotropic Structures... 25
   IV. Determination of Shape.......................... 45
    V. Numerical Density and Mean Caliper Diameter....... 57
   VI. Thickness of a Lamina and Parallel Cylinders........ 83
  VII. Statistical Considerations in Stereology
       (R.L. Scheaffer)............................... 94
 VIII. Anisotropic and Nonhomogenous Tissues and
       Organs, Extension of Objects.................... 114
   IX. Section Thickness and the "Holmes Effect" 121
    X. Automated Methods in Stereology................. 127
   XI. Three-dimensional Reconstruction and Other
       Ancillary Stereological Techniques............... 134
  XII. Applications of Stereological Methods to
       Various Organs 149
 XIII. Potential Future Applications of Stereology.......... 189
Appendix I. Programs for Estimating Mean Caliper Diameter
       From Serial Sections 201
Appendix II. Programs for Estimating Stereological Values of
       Mammalian Heart and Lung.................... 223

Bibliography 291
Subject Index 299

Dedication

The authors dedicate this book to Walter S. Tyler, the master of practical stereology, and to August Hennig, the pioneer of theoretical and applied stereology.

Walter S. Tyler   *August Hennig*

Preface

Living systems, such as entire organisms and their parts—the organs—are built of cells, fibers, sheets, follicles, tubes, spaces and globular masses. These structures are all three-dimensional and continuously moving. To study them, their movements often must be stopped, or they must be killed and hardened by fixation and then cut into thin slices. From these flat slices, the scientist must reconstruct the original, living, three-dimensional form in all of its aspects. One method of approaching this problem is offered by the young science of stereology, a branch of applied mathematics.

Many biologists and medical scientists have expressed the need for a small and practical book that will help them to identify three-dimensional structures from flat sections. Stereology, as a formal branch of applied mathematics, will be just 20 years old when this book is published; however, much work has already been done. There is no single volume to which biomedical laboratory workers can turn for practical guidance. The book *Quantitative Stereology* by *E.E. Underwood* is written for metallurgists and ceramicists. The other book is the learned, two-volume *Stereological Methods* by *E.R. Weibel,* which addresses itself to the same audience as does the present book. Both of these authors presuppose that the reader has had advanced mathematical training. A third publication entitled *Quantitative Microscopy* is Appendix II of the textbook *Histology and Human Microanatomy,* by *Elias, Pauly, and Burns.* The present new book is an expansion of the latter publication. We want to give every person who works with sections a guide that can be easily understood and requires a minimal mathematical background. It is possible to write such a guide because stereology and its mathematical background are indeed very easy to understand.

Stereology deals with shapes, sizes, numbers, orientation in space, and densities. To make these concepts understandable, this new book contains a chapter that defines and describes some geometric shapes and gives a basic introduction to trigonometry. Many persons who must interpret flat images are afraid of stereology because it is a branch of mathematics. One purpose of this book is to free the user from a fear of mathematics. Mathematics, as primitive as that necessary for stereology, is really fun and not at all difficult.

William Thomson, the great English physicist and inventor, better known as Lord Kelvin, is often quoted as saying, "When you can measure what you are speaking about and express it in numbers, you know something about it; but when you cannot express it in numbers, your knowledge is of a meager and unsatisfactory kind; it may be the beginning of knowledge, but you have scarcely, in your thoughts, advanced to the state of science, whatever the matter may be." We do not know the source of this statement but believe it may derive from lecture notes taken by one of Thomson's students. It is a forceful statement, one that a good teacher can make to persuade his audience to pay more attention to numerical accuracy.

Many authors who want to emphasize the quantitative aspect of their subject have repeated this quote again and again. We, too, are presenting a book on the quantitative aspect of our science. But quantification is not the essence of science. First, you must know *what* exists and with what kind of objects you are dealing before you can begin to measure and count them. Quantification is often very useful and can lead to a deeper understanding of natural phenomena. But, it is not the essence of science. You can quantify force, giving it strength in pounds, kilograms or tons, but only by lifting a stone with your muscles can you understand the concept. You can determine the height of Michelangelo's sculpture of David; but this measurement gives you no insight into the magnificence of his work. Nevertheless, we have written a book of morphologic quantification; and as we did so, we learned that stereology is not a science in itself. Rather, it is a method that enables us to attain a deeper understanding of the basic concepts of science that we already know.

During his concluding address to the Fourth International Congress for Stereology in Gaithersburg, Maryland in 1975, *Ervin E. Underwood,* then president of the Society, said (and we must paraphrase his statement because it has not been printed): "As a branch of mathematics, stereology has now been concluded. All of the basic formulae have been derived and published. What we can expect in the future of stereology is refinement and expansion of instrumentation and practical applications." How far was he—one of the fundamental builders and pioneers of stereology—from the truth? It will become obvious to all readers that stereology bristles with unanswered and open questions. We do not refer to new scientific questions that can be answered by known stereological methods. Rather, we mean new problems concerning the organization of space itself. We are confronted with them daily and forced to develop new mathematical methods.

Hans Elias
Dallas M. Hyde

Acknowledgments

We hope to be excused for wording this section in a personal manner. But as there are two co-authors of this book, each of us will have his say in the first person singular. As usual, the senior is permitted to begin.

Early in 1980 my best friend, Dr. John E. Pauly, telephoned to tell me that he had been appointed editor-in-chief of the *American Journal of Anatomy*. He said that he wanted to publish a review article on a timely, widely applicable new field of investigation in each issue of the journal. He invited me to write a comprehensive article on stereology. I agreed, but I asked him for permission to recruit a co-author familiar with the application of computers to stereology, and I proposed Dr. Walter S. Tyler, then Director of the California Regional Primate Center at Davis, to be that co-author. Dr. Pauly agreed, and I asked Dr. Tyler to write a part of that article. Dr. Tyler, however, told me that he intended to resign from the directorship of the Primate Center, since administration outweighed research. In fact, he had already begun to prepare his return to professorship of Veterinary Anatomy at the University of California at Davis. His moving to another position kept him too busy and left him no time for writing. He recommended that I invite his co-worker, Dr. Dallas M. Hyde, to step into the gap. This I did. We have dedicated this book to Walter Tyler because he brought the two of us together. Meanwhile, we have done much other research together.

Dr. Hyde and I worked on the review article on stereology at my home in San Francisco when another and very good friend dropped in. It was Dr. Charles F. Visokay, Medical Editor of the S. Karger Publishing Company. Seeing what we were doing, he suggested that we write a book on stereology for the S. Karger Publishing Company.

I am also grateful to *Dr. Ralf Gander,* scientific director of the Wild-Heerbrungg Optical Company, who caused his firm to donate an M20 microscope to me "in honor" of my retirement from the Chicago Medical School. It is the instrument shown in figure 5.11. All the photomicrographs in this book with a superimposed luminous scale or counting grid are taken with this instrument. This microscope was used also for the production of all the colored photomicrographs in the fourth edition of our (*Elias, Pauly,* and *Burns*) *Histology*. And I use it for much of my research.

The Ernst Leitz Company in Wetzlar also donated an excellent microscope to me.

Finally, I want to thank my co-worker of many years, *Dr. August Hennig* in Munich, for enormous contributions to my work and to stereology in general. *Dr. Melvin M. Grumbach,* chairman of the Pediatrics Department of the University of California, has given me hospitality ever since my guest professorship at the University of Heidelberg ended. He has given me laboratory and office space in his department and the permission to use department equipment. Also, *Dr. Joseph R. Goodman,* chief of the electromicroscopy laboratories at the Pediatrics Department has assisted me with space, active help and permission to use his equipment.

Dr. Edward A. Smuckler, chief of the Pathology Department, has given me continuous access to the collections of the Pathology Department. All the material described at the end of Chapter 12 is from his collections.

This list does not exhaust the enumeration of all persons who have helped us. Nobody can work in a vacuum; a multitude of kind friends and helpers has been needed. And we herewith thank all of them including those whom we failed to mention.

Now I turn the pen over to my new friend and co-worker, Dallas Hyde.

Hans Elias

To Dr. Walter Tyler, my mentor and colleague, I join Dr. Elias in dedicating this book to him. Dr. Tyler has provided encouragement and advice throughout my academic career. Dr. Richard L. Scheaffer has made many important contributions to Stereology during his academic career and I am particularly grateful to him for his excellent chapter (7): *Statistical Considerations in Stereology.* I thank Mr. Edward Reus, Ms. Nancy Tyler and Mr. James Inderbitzen for writing the programs of Chapters 11 and 12. Drs. Tyler, Daryl Buss and Charles Plopper are thanked for their review of and comments on Chapter 12. I am most grateful to Dr. Elias for his fatherly attitude in teaching me stereology and giving me this opportunity to collaborate with him on a subject that we both immensely enjoy.

A special note of thanks goes to Trudi Schuster for typing and proofreading this book. I am indebted to Dr. George Cardinet, my department chairman, for his personal encouragement and support which were essential in completing this work. Thanks also go to Ms. Mary Stovall and Ms. Alison Faulkener for photographic and library assistance.

Finally, I thank my wife, Dr. Leigh West Hyde, for her patience and encouragement throughout this endeavor.

Dallas M. Hyde

I. Some Elementary Geometric Concepts

Introduction

We want this book to be very useful. A set of rules can be useful without being understood, like the directions to prepare a staining solution. But if a technician has knowledge of chemistry these directions acquire meaning, and the whole labor of staining, instead of being a boring routine, can be fun.

This chapter will introduce you to a few basic concepts of geometry. Readers trained in mathematics may skip this chapter, but even for them, reading it could be useful because it presents very simple explanations of basic concepts, which may be useful when presenting these concepts to beginners in stereology.

Our first forays into the field of stereology were intuitive, founded on common sense attempts to answer the question, "What could the three-dimensional structure of this thing be, of which I see only an extremely thin slice?" We were, then, confronted with images which were very difficult to interpret. To illustrate, we will describe an approach used by *Vincent Hall* (1955) when attempting to visualize the three-dimensional shape of podocytes in the kidney (fig. 1.1). Hall's approach was as follows: he was confronted with hundreds of flat images which resulted from cutting complicated three-dimensional objects (all similar to each other but not entirely alike). To unravel the puzzle Hall asked himself, "What kind of object could yield sections that look like these?" So he built a strange object of modeling clay and sectioned it with a knife. Repeating this process several times, he arrived by trial and error at a three-dimensional model of a podocyte that, when cut, would yield flat images very similar to those seen in hundreds of electron micrographs. Such a three-dimensional image does not need to be made of clay. It can be a drawing showing a view of the object as if seen from a certain angle. Such a drawing is called a *stereogram*. Stereograms are products of experience and imagination. To draw a valid stereogram like that in figure 1.2, one must have seen a great number of sections of such objects at various magnifications and one must be able to imagine shapes which one has never seen. To be truthful, this stereogram (fig. 1.2) is not by Hall, but was produced by us 5 years later and again modified in 1978 to account for more recently discovered detail.

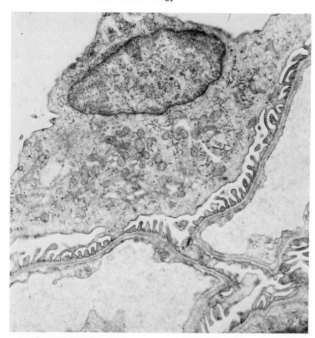

Fig. 1.1. Electron micrographs of renal podocytes.

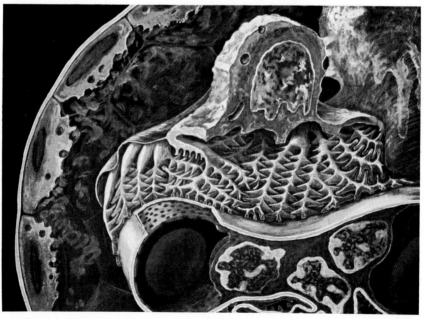

Fig. 1.2. Stereogram of podocyte. [Reproduced with permission from Elias, H.; Pauly J.E., and Burns, E.R.: Histology and human microanatomy; 4th ed. (Wiley, N.Y. 1978).]

Some Elementary Geometric Concepts

A *stereogram*, then, is a conceptual, *generalized* visualization drawn in perspective of a three-dimensional object. A *reconstruction* from serial sections may look very similar, but portrays one *specific* object and its origin is very different. It is based on stacked images of successive serial sections. Methods of reconstruction will be discussed in Chapter 11.

In its early days stereology was called the "geometry of sectioning." It is, therefore, appropriate to talk a little about geometry, specifically about shapes one encounters in histology. We shall begin with very simple two-dimensional shapes. (We can disregard, for the moment, the thickness of the "section" or slice, because we always look at it through the microscope from above, and hence the image appears flat.)

The Circle

A circle is the locus of all points on a plane which have the same distance from one point, the center. This constant distance is called the radius (r) (fig. 1.3). A circle is perfectly round; in fact it was considered the perfect shape by early astronomers. When Kepler noticed that the planets move in orbits that are not perfect circles, but "merely" ellipses, he was distraught and began to doubt the existence of God.

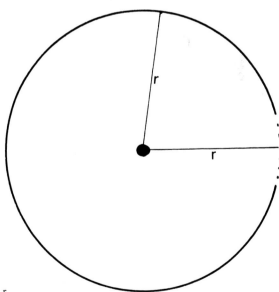

Fig. 1.3. Circle, Radius r.

The Ellipse

The ellipse is an oval figure of great precision. Although the name oval derives from *ovum* (Latin, egg), the ellipse, unlike a chicken's egg, has two ends of exactly equal curvature and two sides with curvatures that are exactly alike. The ellipse has a center (C) and two focal points (F). Its shape (whether thick or slender) can be defined by the eccentricity, the quotient of the distance between the two focal points divided by the length (major axis) of the ellipse (fig. 1.4). An ellipse can be defined as the locus of all points for which the sum of the distances from both focal points is the same. Therefore, an easy method for drawing an ellipse is to attach a string with thumbtacks to both focal points, stretch that string with a pencil, and draw while moving the pencil so that the string remains taut (fig. 1.4). This string is the simplest ellipsograph. The points are called focal points from *focus* = fire because a lit candle placed in one focal point of an elliptical chrome-lined box, will light a match held in the other focal point (fig. 1.5). Since the angles between the two segments of the string and tangent (represented by the ruler in fig. 1.4) are equal, all rays issuing from one focal point are reflected through the other.

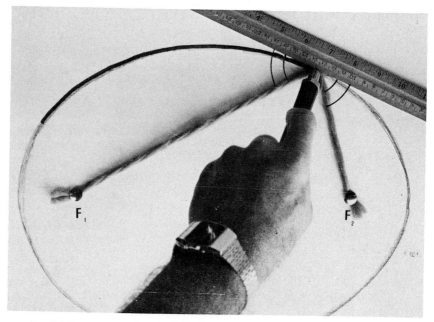

Fig. 1.4. Generating an ellipse with string and pencil.

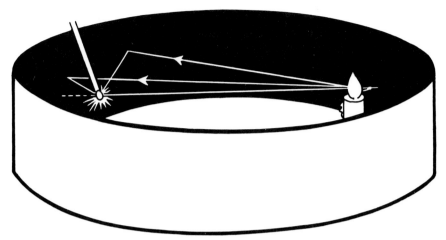

Fig. 1.5. Demonstration of flame "ghost" at focal point of ellipse.

A few buildings are constructed in the shape of an ellipse. They are called whispering galleries. If the walls are smooth, a person standing in one focal point can hear the faintest whispering of another person standing in the other focal point. The most famous whispering gallery was the Ear of Dionysos at Syracuse in Sicily. It was an artificial cave, a huge prison vault according to mythology, shaped like an ellipse. The prisoners were confined to an area near one focal point, while a secret agent, hidden from the prisoners by a wall, was placed at the other focal point. This is how Dionysos was kept informed about conversations among prisoners. Ruins of this cave still exist, but the alleged acoustic principle of the device is no longer apparent. Among contemporary whispering galleries of this kind are the Hall of Karyatids in the Louvre in Paris, the entrance hall in the Palace of the Bishop in Würzburg and a hall in the Castello Sforza in Milan.

While sophisticated mathematicians use the eccentricity of the ellipse to identify its shape, stereologists prefer the axial ratio Q—the quotient of length to width—because this can be directly measured. In one kind of ellipse, the axial ratio of length to width is 1. This is the circle. Its eccentricity equals zero. A circle can easily be drawn with the string ellipsograph by fastening both ends of the string to one thumbtack.

Cylinders

A cylinder is the locus of all straight lines parallel to each other which pass through a curved line (fig. 1.6). The parallel lines are called the generators

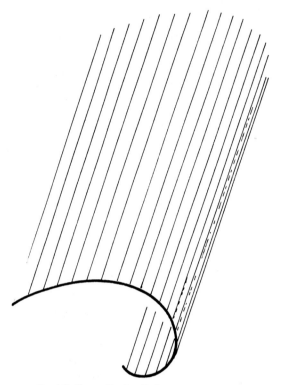

Fig. 1.6. Generalized cylinder.

of the cylinder while the single curved line is called the directrix. If the directrix is a closed polygon, the cylinder is a prism, such as cells of simple columnar epithelia and most skeletal muscle fibers where closely packed. If the directrix is a circle and the generators are perpendicular to the plane of the circle, the resulting three-dimensional formation is called a right circular cylinder. Most peripheral nerves have this shape, as do many arteries and isolated skeletal muscle fibers such as those of some facial muscles and those of the external sphincter ani. If the directrix of a cylinder is an ellipse, the cylinder is elliptic. Veins and the *Taeniae coli* have such shapes.

Ellipsoids

There are two basic types of ellipsoids: rotatory ellipsoids and triaxial ellipsoids. Any ellipsoid, when freely suspended in space, casts an elliptical

Some Elementary Geometric Concepts

shadow. When an ellipsoid is cut at any angle, the sections have the shape of ellipses.

A rotatory ellipsoid can be generated by rotating an ellipse around one of its axes. Cut an ellipse out of a piece of cardboard; use a template or string (fig. 1.4) to draw the ellipse. Paste a string along its major (the longer) (fig. 1.7a) axis. Then hold the string at both ends and spin it. The result will be a *prolate* rotatory ellipsoid. Many nuclei of columnar epithelium and many fibroblast nuclei approach this shape; so do grains of rice. Any section perpendicular to the axis of rotation will be a circle. Oblique sections will be ellipses of various shapes. If the string is fastened along the minor axis and spun, the result will be a lens-shaped object—an *oblate* rotatory ellipsoid (fig. 1.7b). Also, in this case sections perpendicular to the axis of rotation will be circles; all other sections will be more or less oblong ellipses. The most oblong ellipses result from cutting parallel to the axis of rotation. Nuclei of the upper layers of a stratified, squamous epithelium usually are oblate ellipsoids. A sphere is a rotatory ellipsoid that can be generated by spinning a circle around any of its diameters.

Some ellipsoids cannot be produced by rotation. They have three unequal axes perpendicular to each other. Their shadows or sections are ellipses of various shapes. There is a unique cutting angle for any triaxial ellipsoid that

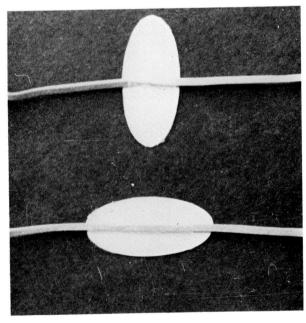

Fig. 1.7. Generating a rotatory ellipsoid from a cardboard ellipse and a string: (*a*) prolate ellipsoid and (*b*) oblate ellipsoid.

will yield a circle. Endothelial nuclei and nuclei of pulmonary type I epithelial cells are examples of triaxial ellipsoids.

Isodiametric Versus Anisodiametric Objects

Many objects are longer than they are wide and wider than they are thick. But the sphere has only *equal* diameters. It is *the* ideal isodiametric (ἰσός Greek, equal) object. However, any object whose diameters do not differ very much from each other can be called isodiametric, for example a cube. The word diameter has many definitions. In stereology, it means a caliper diameter, i.e., the distance between two opposite parallel tangent planes (fig. 1.8). An isodiametric object has a caliper diameter that is constant or varies only slightly. For a cube of edge length 1, the caliper diameter varies between 1.0 and 1.5537. Still, these extreme diameters are close enough so that a cube can be said to be isodiametric. On the other hand, consider a yardstick $\frac{1}{8}$ inch thick, $1\frac{1}{8}$ inches wide and 36 inches long. Its shortest diameter is $\frac{1}{8}$ inch; its longest diameter (when the yardstick is standing on a corner with the opposite corner vertically above it) is slightly greater than $36\frac{1}{8}$ inches, that is, about 289 times longer than thick. This is certainly an anisodiametric object. A renal glomerulus, which is usually shaped like a slightly flattened spheroid, is classified as isodiametric. The division of objects into isodiametric and anisodiametric is rather loose, the sphere being the only truly isodiametric object. Yet the tendency is to call a human neurocranium isodiametric and that of a fox anisodiametric. Although these two words are rather vague, they are used frequently.

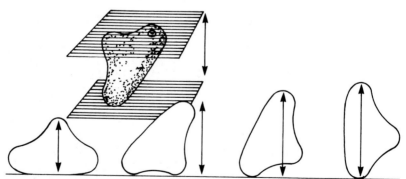

Fig. 1.8. The mean caliper diameter of a solid is estimated by taking multiple measurements of its height in all possible positions.

Some Elementary Geometric Concepts

Some interesting shapes encountered in histology are polyhedra; the simplest are prisms represented by cells of simple cuboidal and columnar epithelia and by closely packed skeletal muscle fibers. More complicated shapes are found in cells in the interior of tissues, such as fat cells in adipose tissue and the stratum spinosum of the epidermis. They acquire, by compression of adjacent cells, the average shape of tetrakaidekahedra (14-faceted bodies). This is explained in detail elsewhere (*Elias et al.* 1978, p. 44).

A Little Trigonometry

Trigonometry plays an important role in the derivation of stereological formulas, and trigonometric functions even appear in some of them. Trigonometry is a very simple and easy science, yet many people are afraid of it because it is usually presented in a very clumsy fashion in school. Therefore, in the following section visual images of trigonometric functions have been created. In fact, the names of the functions are based on the visual images. Even without practice, the descriptions will permit any reader—no matter how primitive his high school trigonometry has been—to *see* a sine and tangent grow and a cosine and cotangent shrink as the angle increases. The limits of these functions also become naturally visible.

In every handbook of formulas and tables, are found the classic definitions of trigonometric functions based on a right triangle (fig. 1.9) in which the side opposite the right angle is called the hypotenuse (h); the angle α

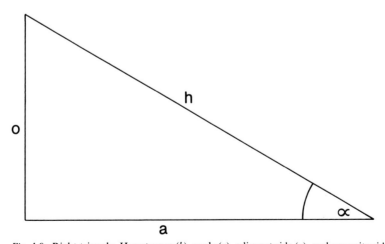

Fig. 1.9. Right triangle. Hypotenuse (h), angle (α), adjacent side (a), and opposite side (o).

(to the right in the diagram) has an adjacent side (*a*) and an opposite side (*o*). The functions are conventionally written as follows:

sin α = o/h (sine)   cos α = a/h (cosine)
tan α = o/a (tangent)   cot α = a/o (cotangent)
sec α = h/a (secant)   csc α = h/o (cosecant)

Every angle has a "complementary" angle, which is 90° − α, from which word the prefix "co-" derives. If the sum of angles in any plane triangle equals 180° and 90° of this 180° are already taken up by the right angle opposite the hypotenuse, obviously, when α is on one side of the hypotenuse, the remaining acute angle is 90° − α, i.e., the complementary angle to α. Hence, we have the following important relationships: cos α = sin (90° − α), cot α = tan (90° − α) and csc α = sec (90° − α).

All of these relationships can be *visualized* by inscribing a right triangle in a unit circle; i.e., a circle of radius 1 (fig. 1.10). Using this circle, the names for the trigonometric functions acquire sense and their ranges and limits

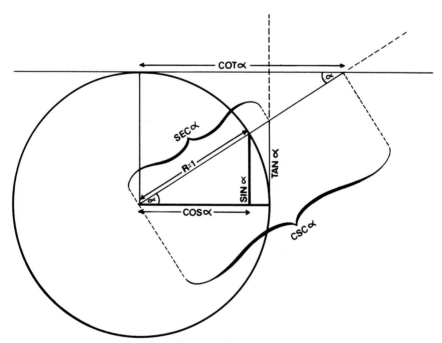

Fig. 1.10. Right triangle superimposed on unit circle to illustrate sin, cos, tan, sec, csc, and cot functions.

Some Elementary Geometric Concepts

become immediately apparent. The angle α is formed by a horizontal radius of the circle and a freely moving leg. Originally, the word sine meant the chord (sinew) of a bow. But in trigonometry we use only half a chord for the sine and half a bow for its arc. Using the conventional definitions given above, we can see the sine of α directly. It is the quotient of the opposite side divided by the hypotenuse and is visible as the vertical line reaching from the point where the free leg of the angle intersects the circle. This hypotenuse is the free or movable radius of the circle, which equals 1. Let us repeat: the vertical line from the point where the free leg of α intersects the circle *is* its sine. A simple way to observe the waxing and waning of the sine in the first two quadrants of a circle is to suspend a weight by a string at the rim of a wheel and hide the lower half of the wheel behind a board. The free limb of the angle is represented by the spoke leading to the suspension point. This is shown in figure 1.11. We can also see that for α = 0, sin α = 0 and that as α grows, sin α grows, but it can never be greater than 1,

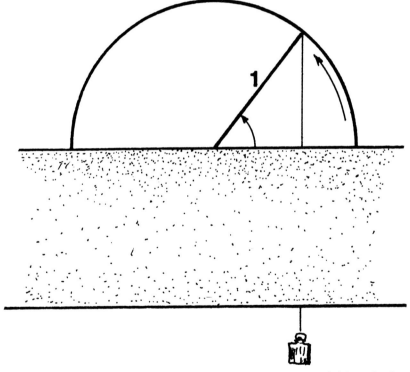

Fig. 1.11. Generating a sine curve using a circle and a weight suspended from the rim of the circle.

which happens when α = 90°. As soon as α is greater than 90°, sin α becomes shorter again, and when α = 180°, sin α is again zero. When α becomes greater than 180°, its free leg will point downward and the sine of α will become negative (hidden in fig. 1.11); hence for α = 270°, sin α = −1.

If we now plot all the values of α on a horizontal line (abscissa) of infinite length and the corresponding values of sin α as the ordinates, then we will obtain a wavy line called a sine curve, such as is often used to represent an alternating electrical current (fig. 1.12).

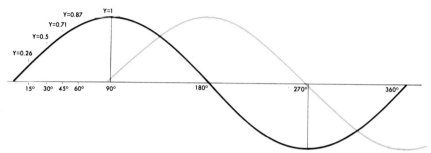

Fig. 1.12. Sine curve. Values for sin vary from 0 to 360. The angles are plotted on the abscissa (horizontal axis). The value of the sine for each angle is plotted as the ordinate.

The cosine, represented in figure 1.10 as a double arrow, behaves very much like the sine. But it equals 1 when α = 0 and 0 when α = 90°. A cosine curve intertwines with a sine curve. It looks just like a sine curve; but at α = 0, it equals 1 and at α = 90°, it is 0. The phase difference is 90°.

The word tangent comes from the Latin *tangere*, to touch. Again, refer to figure 1.10. A vertical line *touching* the unit circle on the right side is intersected by the continuation of the free leg of α. When α = 0, tan α = 0; when α = 45°, tan α = 1. When α increases beyond 45°, tan α also increases until at α = 90°, tan α = ∞.

The cotangent is the tangent of the complementary angle to α. It is represented by a horizontal line touching the unit circle above in figure 1.10. Cot α is the length of that line from the top of the circle to its intersection with the free leg of α. With just the slightest imagination you can see that when α = 0, cot α = ∞, and when α = 90°, cot α = 0.

The secant (from the Latin *secare*, to cut) is the length of the free leg of α to its intersection with the tangent line; and the cosecant of α is its length from the center of the circle to its intersection with the cotangent line.

Some Elementary Geometric Concepts 13

(Again, refer to figure 1.10). Now you can figure out for yourself the limits of the secant and of the cosecant, just by looking at figure 1.10 and imagining the free leg of α swinging about the center.

The changing signs of trigonometric functions can be summarized according to quadrants: quadrant $1 = 0° - 90°$; $2 = 90° - 180°$; $3 = 180° - 270°$; and $4 = 270° - 360°$. Lines that go up or to the right have a positive sign, those going down and to the left are negative (table 1.1).

Table 1.1 Changing signs of trigonometric functions according to quadrants

Quadrant	sin	cos	tan	cot	sec	csc
1	+	+	+	+	+	+
2	+	−	−	−	−	+
3	−	−	+	+	−	−
4	−	+	−	−	+	−

There are two methods of measuring angular magnitude: (a) In the sexagesimal system, an angle generated in one complete rotation is divided into 360 equal parts—*degrees*. Each degree is subdivided into 60 equal parts—*minutes*—which in turn are subdivided into 60 equal parts—*seconds*. (b) In the circular system, the unit angle is the angle subtended at the center of a circle by an arc equal in length to the radius of the circle, a *radian*. Since the circumference of one complete circle is $2\pi r$ or 2π radian by definition, corresponding to an angle of 360°, we can calculate the angle subtended by a radian. It is $360/2\pi = 360°/6.2832 = 57.29°$. In higher mathematics radian measurement is frequently used, as is exemplified by the mean caliper diameter estimation from a serially reconstructed image in Chapter 11.

Table 1.2 summarizes common trigonometric values in the first quadrant according to radians and degrees.

Table 1.2 Common trigonometric values in the first quadrant according to radians and degrees

Arc in radian	Angles in degrees	sin	cos	tan	cot	sec	csc
0	0	0	1	0	∞	1	∞
$\pi/6$	30	1/2	$\sqrt{3}/2$	$1/\sqrt{3}$	$\sqrt{3}$	$2/\sqrt{3}$	2
$\pi/4$	45	$1/\sqrt{2}$	$1/\sqrt{2}$	1	1	$\sqrt{2}$	$\sqrt{2}$
$\pi/3$	60	$\sqrt{3}/2$	1/2	$\sqrt{3}$	$1/\sqrt{3}$	2	$2/\sqrt{3}$
$\pi/2$	90	1	0	∞	0	∞	1

The Concept of Curvature

An irregularly curved line on a plane exhibits a crookedness that varies from place to place. The degree of curvature can be identified by drawing a tangent circle to the curve at any location (fig. 1.13). The radius, r, of the tangent circle is the local radius of curvature. Curvature itself is the degree of crookedness and precisely the reciprocal of the radius of curvature (c) or

$$c = \frac{1}{r} \qquad (1.1)$$

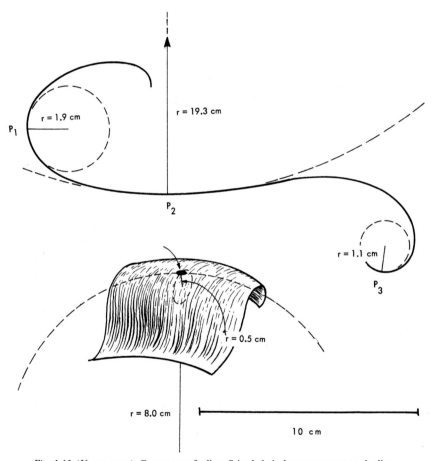

Fig. 1.13. (*Upper curve*). Curvature of a line. Stippled circles are tangent to the line.

Fig. 1.14. (*Lower curve*). Curvature of a surface. Stippled circles are tangent to the surface.

Some Elementary Geometric Concepts

The entire curve can be described by its mean curvature, the geometric mean of the curvatures for each location of the curve.

An irregularly curved surface in space is more crooked in one direction than in another at any one location (fig. 1.14). In space, the curvature (K) of a surface at any one location is the reciprocal of the product of both radii of curvature tangent to the surface, so that

$$K = \frac{1}{r_1 \cdot r_2} \qquad (1.2)$$

This is called Gaussian curvature. In this manner, when one of the radii of curvature is negative, K becomes negative automatically.

This means that a surface is curved in such a way that the center of one tangent circle lies on one side of the surface and that of the other on the opposite side. To surfaces of this kind belong the saddle surfaces, like a rider's saddle. The articular surfaces between the trapezoid bone of the human wrist and the first metacarpal bone are saddle surfaces. These permit circumduction of the thumb without allowing rotation (fig. 1.15). The curvature of a surface at a saddle-shaped location is also calculated by formula 1.2, where one of the r's is negative. Mean surface curvature ($\bar{K}$) is the average of curvatures for all points of an irregularly curved surface.

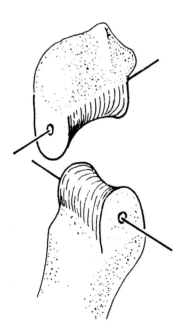

Fig. 1.15. Schematic representation of the saddle joint between the wrist and the thumb, trapezium and the first metacarpal.

II. Basic Definitions, Terminology, and Symbolism

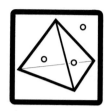

Introduction

On May 11, 1961, the word *stereology* was coined and the International Society for Stereology founded on the Feldberg. The emblem of the society, shown above, was designed during this meeting. On the following day, the Society was incorporated as a nonprofit organization at Neustadt, Schwarzwald, Germany (*Bach* 1963a, *Elias* 1977). This marked the establishment of stereology as a formal science. Subsequent registrations of the Society, took place in Vienna (1963) and in Chicago (1967) and served to give it a broad, international basis.

Stereology is three-dimensional interpretation of flat images, such as sections and projections, by criteria of geometric probability. The earliest stereologists were astronomers like Aristarchos, Ptolemaeus, Copernicus, Brahe, and Kepler, who interpreted the apparent motions and magnitudinal changes of planets as projected on the inner surface of the hollow celestial sphere in terms of movements in three-dimensional space. Stereology's chief users today are histologists, materials scientists, petrographers, physiologists, and ceramicists, who must obtain information about the spatial organization of organs, rocks, and materials from true sections, such as the plane of polish in metallography, or thin slices in histology and geology. Also, contemporary astronomers and cosmologists practice stereology. But for them, the interpretation of projected images in terms of depth is a matter of course. They think stereologically all of the time, without mentioning the word.

Our own, simple, down-to-earth stereology is practiced chiefly by superimposing line and point patterns on flat images by which the sectional images of component parts can be quantified. Counted points and measured line segments are entered into basic formulas. These formulas, as will be seen below, are quite simple and straightforward. Similarly, most of their derivations have been straightforward and easily developed. Some mathematicians, however, have unnecessarily employed highly sophisticated analytic methods to obtain the same results.

This chapter is intended as a practical guide to stereologic concepts and techniques that can be used by morphologists and technicians with little mathematical background. Our focus is on practical applications. Nevertheless, derivations of most basic formulas will be given, just to show how easy it is.

Originally, most basic formulas were derived independently by several different investigators confronted with similar problems. This polyphyletic origin of stereology is the reason that, 19 years after the discipline was founded, no formal common terminology and symbolism yet exist. Two main systems coexist at this time: the system we use in this outline is that of DeHoff, Rhines, and Underwood, metallurgical scientists. The other system, which is not used herein, is that by Weibel, a medical scientist.

Basic Definitions

The words *morphometry* and *stereometry* are often used synonymously with *stereology*. In reality, they refer to three different areas of mathematics.

Morphometry is the measurement of structures. A tailor who measures someone for a suit practices morphometry. The focusing devices of many photographic cameras that measure distance are morphometric instruments; so are the odometers in automobiles. However, morphometry is often the aim of stereology. In morphometry the questions asked are: "How large?" and "How many?" Stereology is often the only method by which answers to these questions can be obtained.

Stereometry is the same as solid geometry. It is that part of mathematics that calculates volumes and surface areas of exactly defined solids such as cubes, spheres, and cones. But this word is used in the sense of stereology by *Saltykov* (1958).

Stereology is the three-dimensional interpretation of flat images (sections and projections) by criteria of geometric probability. Stereology can also be defined as extrapolation from two-dimensional to three-dimensional space. Although a branch of applied mathematics, it is an easy science to learn. Its rules and formulas are simple and straightforward. They are not approximations, but are mathematically precise. Stereology deals with points of zero dimensions (P), lines (straight and curved) and their one-dimensional lengths (L), areas of two dimensions (A or S), and volumes of three dimensions (V).

A point P may exist as a dimensionless spot in space, like the center of a renal glomerulus, a fat cell, or a lysosome. It may be a test point used for measurements. It can be generated by the intersection of two lines on a surface, by the intersection of a line with a surface in space, or by the intersection of three surfaces in space. It can also be the center of a profile (fig. 2.1).

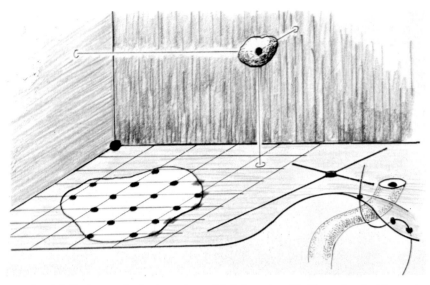

Fig. 2.1. Various methods of creating and defining points (P). [Reproduced with permission from Elias, H.; Pauly, J.E., and Burns, E.R.: Histology and human microanatomy; 4th ed. (Wiley, N.Y. 1978).]

A line of length L may be straight or curved in space, such as the axis of a glandular tubule or of a skeletal muscle fiber, or the total length of all capillaries in a large organ. It may be a test line or the trace of a membrane, i.e., its intersection with the cutting plane (fig. 2.2).

The letter A is used in stereology for a test area within which features are counted, such as a square or a circle of known size (fig. 2.3). S signifies the area of a curved surface in space, such as the skin of a potato (fig. 2.4).

The number of features counted in a test area is given by the letter n or P (fig. 2.3), while the number of tissue components, such as pancreatic islands, nuclei, or microbodies in a volume, is given by the letter N. Of these three (n, N and P), we prefer to use P, since the number of particles in a volume N_V is the same as the number of their centers of gravity P_V in that volume. In some instances, for the sake of convenience, we also use n and N. In any case, the reader will notice from the context what is meant.

The volume V may be a unit volume, such as a cubic millimeter; it may be a total volume, such as the volume of the entire kidney; or it may be a volume fraction occupied by a component of an organ, such as the collective volume of all glomeruli in an entire kidney or of medulla compared with cortex. It may also be an average volume, for example, the average volume of nuclei, mitochondria, or lysosomes in a cell or tissue.

Fig. 2.2. (*Top*) Various meanings of the terms line (*L*) and length (*l*). [Reproduced with permission from Elias, H.; Pauly, J.E., and Burns, E.R.: Histology and human microanatomy; 4th ed. (Wiley, N.Y. 1978).]

Fig. 2.3. (*Bottom*) The use of a test area A to count profiles. [Reproduced with permission from Elias, H.; Pauly, J.E., and Burns, E.R.: Histology and human microanatomy; 4th ed. (Wiley, N.Y. 1978).]

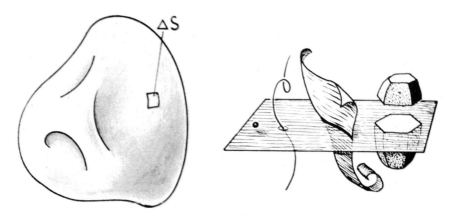

Fig. 2.4. (*Left*) A curved surface. [Reproduced with permission from Elias, H.; Pauly, J.E., and Burns, E.R.: Histology and human microanatomy; 4th ed. (Wiley, N.Y. 1978).]

Fig. 2.5. (*Right*) The principles of dimensional reduction. [Reproduced with permission from Elias, H.; Pauly, J.E., and Burns, E.R.: Histology and human microanatomy; 4th ed. (Wiley, N.Y. 1978).]

Since, in stereology, we extrapolate from two- to three-dimensional space and take all measurements on plane surfaces, test volumes do not exist as operational devices. Many authors like to use subscripts to specify the chief symbols. However, in most cases, the context within an equation adequately identifies the meaning of a letter. On the other hand, in equations involving structures of different kinds, we prefer to spell out the subscripts. This makes a symbol quite explicit.

The Law of Dimensional Reduction

A true section, without thickness, through an n-dimensional object produces an $(n - 1)$-dimensional profile. Thus, a point that has no dimension will have practically no chance at all of being cut by a section. But a granule may be caught between the two cutting surfaces of a slice. A line representing a fiber, for example, yields a dot (point of zero dimensions) when cut by a plane. A surface, such as an extremely thin membrane, yields a one-dimensional line on section. A solid, like a sympathetic ganglion, a cell, a peroxisome, or a pebble has an area as its profile. Conversely, an n-dimensional figure in a section is, in general, a profile of an $(n + 1)$-dimensional structure. (These relationships are illustrated in fig. 2.5).

Terminology

Section: a plane without thickness generated by one cut through a solid (fig. 2.6).

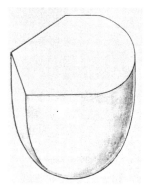

Fig. 2.6. A section. [Reproduced with permission from Elias, H.; Pauly, J.E., and Burns, E.R.: Histology and human microanatomy; 4th ed. (Wiley, N.Y. 1978).]

Slice: a histologic or ultrathin "section" of finite thickness t generated by two cuts (fig. 2.7); also a thin ground slice through a rock, as used in geology, just as a slice of ground bone or tooth. Underwood calls it a foil, but excludes foils created by pressure.

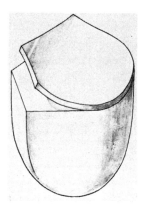

Fig. 2.7. A slice. [Reproduced with permission from Elias, H.; Pauly, J.E., and Burns, E.R.: Histology and human microanatomy; 4th ed. (Wiley, N.Y. 1978).]

Slab: a portion of an object between two specified cutting planes. A slab may be opaque or translucent; in gross anatomy it can be as much as 1-cm thick, like the slices of brain used in neuroanatomy courses. A slab can also be a so-called section of computer-aided tomography (Chapter 11). A slab may consist of discarded thin slices between two cutting planes. The distance between the two preserved cutting planes is likewise denoted by the letter t (fig. 2.8).

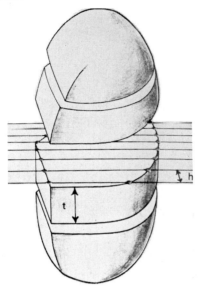

Fig. 2.8. Slabs and intersections. [Reproduced with permission from Elias, H.; Pauly, J.E., and Burns, E.R.: Histology and human microanatomy; 4th ed. (Wiley, N.Y. 1978).]

Profile: a section or slice through an individual component of a solid (fig. 2.9).

Fig. 2.9. Profiles in a plane of section (*left*) and in a slice (*right*). [Reproduced with permission from Elias, H.; Pauly, J.E., and Burns, E.R.: Histology and human microanatomy; 4th ed. (Wiley, N.Y. 1978).]

Trace: a section of a two-dimensional surface, such as a membrane, or interface (boundary between two components of a space-filling aggregate). A trace appears as a line (fig. 2.10).

Intercept: the length of that segment of a test line, which falls on the area of a profile (fig. 2.11).

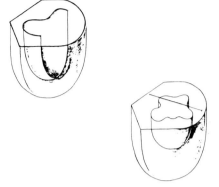

Fig. 2.10. (Left) Traces. [Reproduced with permission from Elias, H.; Pauly, J.E., and Burns, E.R.: Histology and human microanatomy; 4th ed. (Wiley, N.Y. 1978).]

Fig. 2.11. (Right) An intercept. [Reproduced with permission from Elias, H.; Pauly, J.E., and Burns, E.R.: Histology and human microanatomy; 4th ed. (Wiley, N.Y. 1978).]

Intersection: the point at which a test line crosses a trace (fig. 2.12; also the dots in fig. 2.8).

Axial ratio: the quotient of length over width of a profile (fig. 2.13).

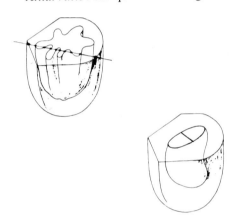

Fig. 2.12. (Left) Intersections. [Reproduced with permission from Elias, H.; Pauly, J.E., and Burns, E.R.: Histology and human microanatomy; 4th ed. (Wiley, N.Y. 1978).]

Fig. 2.13. (Right) Axial ratio (length/width). [Reproduced with permission from Elias, H.; Pauly, J.E., and Burns, E.R.: Histology and human microanatomy; 4th ed. (Wiley, N.Y. 1978).]

Symbols

Because of the cumbersome nature of words in formulas, a standard set of symbols has proved useful in stereology. Below is a summary of the De Hoff-Rhines-Underwood system, which we hope will be universally accepted; Weibel's symbols are listed in parentheses, as are suggested improvements of the DeHoff-Rhines-Underwood system.

P number of points, regardless of their origin
P_T number of test points
P_n number of points which hit a structure
P_P point fraction (P_n/P_T)

P_L number of intersections per unit length of test line (I, *Weibel*)
P_A number of points per test area. (When these points are the centers of profiles of a linear tissue component, such as kidney tubules, *Weibel* uses the symbol Q.)
P_V number of points per unit volume (also notated as N_V)
ℓ length of lineal element ($L = \sum \ell$)
L_L lineal fraction; length of intercept per unit length of test line
L_A length of lineal elements per unit test area (B_A, *Weibel*)
L_V length of lineal elements per unit volume (to visualize this concept, stuff a long string into a tiny, transparent box)
A test area or planar area of intercepted features
S area of curved or plane surface or interface in space
A_A area fraction
S_V surface area per unit volume. (To imagine this visually, take a small, transparent box of known volume V and stuff into it a crumpled sheet of paper of known area S. A surprisingly large sheet of thin paper can fit into a tiny box.)
V volume
V_V volume fraction (for example, the volume of raisins within a cake)
N number of particles
N_V number of particles per unit volume (perhaps better, P_V) (for example, the number of raisins in a cube cut out of that cake)
n number of profiles
n_A number of profiles per unit test area (perhaps better, P_A; N_A, *Weibel*)
Q axial ratio (length/width). In Weibel's system, Q means something else (see above).
t thickness of slice or slab
T thickness of a sheet in space (τ, *Weibel*)
F index of folding or wrinkling
K mean surface curvature
D diameter (also H or h, if caliper diameter is meant) of a particle
d diameter of a profile
I internodal length = distance between points of branching
Ω degree of orientation

III. Stereological Measurements of Isotropic Structures

Introduction

The chief aim of stereology is to determine the following from sections and slices: three-dimensional shapes, orientation in space, volumes and volume ratios, curved surface areas, curvature of surfaces, lengths of curved linear elements, size distributions of corpuscles or organelles, and number of particles per unit volume. We shall proceed in the order of difficulty of understanding, beginning with the easiest (volumes), then progressing to more difficult stereologic problems (the estimation of three-dimensional shapes and the number per unit volume).

It is assumed in most phases of stereology that an organ, a rock, or an alloy is *isotropic*, with its components randomly arranged in space. The stereologic methods that follow apply to organs of isotropic construction or to random samples. Determinations of volumes and volume ratios are independent of arrangement. However, since anatomists often deal with organs that are anisotropic in structure, with oriented arrangement of parts or gradients of density, stereologic techniques for dealing with such organs will be covered in subsequent sections of this book.

Stereology is practiced by measuring and counting points and profiles in sections. For most stereological problems, this is done by superimposing grids of lines or squares and circles over an image. It can be accomplished in many different ways: A glass disk with an engraved pattern can be inserted into the eyepiece; a pattern can be reflected into the path of vision by a camera lucida; or the slide can be projected on a screen or a paper on which the desired pattern is drawn. A pattern engraved on a sheet of plexiglass can be superimposed on a photographic print of the object; this method is frequently used in electron microscopy.

The choice of a specific pattern is a matter of taste, but may also depend on the kind of material examined. Figure 3.1 shows a few such patterns. Test patterns with many points, lines, and fields, sometimes a few hundred, are often used by electron microscopists because of the minute details in their pictures. While test patterns *a–e* in figure 3.1 employ squares as their basic shape there are others that emphasize the equilateral triangle (*f–h*) or semicircles (*i*).

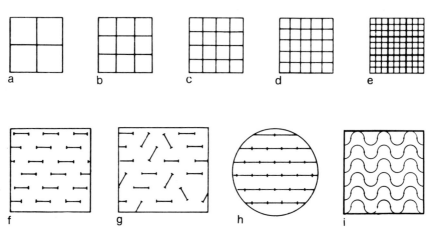

Fig. 3.1. Various test patterns or grids.

It has been shown experimentally that greater efficiency (that is, equal accuracy in less time) is achieved by displacing a simple pattern frequently and systematically to new positions. This is due to the fact that a simple pattern can be scanned faster than a complicated one by a human observer. Even more efficient is the use of a simple pattern with systematic displacement of the specimen. In light microscopy, the specimen can be systematically displaced using a mechanical stage with click-stops (a sampling stage), which eliminates bias, making it impossible for the observer to focus on a point of interest. Independence from human bias is very important. In the absence of a sampling stage, one can guard against bias by moving the specimen approximately equal distances without looking into the microscope. The writers prefer pattern *c* in figure 3.1, since it is coarse enough for easy observation and its 25 points facilitate expressing the result in percentages.

Unbiased sampling in electron microscopy is usually achieved by mincing fixed tissue into tiny blocks, making planned orientation improbable. After the tissue is sectioned, the observer should not merely photograph areas of special interest, but should use some systematic method of selection. For example, photographs should be taken in the upper left-hand corner of every square formed by grid bars, if a section is located there (fig. 3.2; *Weibel* 1963).

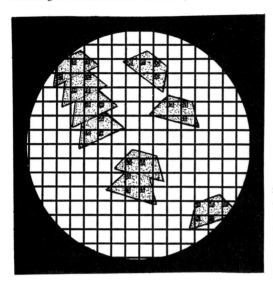

Fig. 3.2. Avoiding bias in TEM by taking pictures only from upper left corners of grid squares. [Reproduced with permission from Elias, H. and Hyde, D.M.: Elementary stereology. Am. J. Anat. *159:* 411–446 (1980).]

Volumes and Volume Ratios

Direct Volume Determination

The easiest method of determining the volume of an object is the immersion method (fig. 3.3). A container of water is placed on a scale and weighed. Then the object, suspended by a thin thread or wire, is immersed until it is fully covered by water but does not touch the bottom of the container. The new weight in grams, minus the weight of the container and water, equals the volume of the object in cubic centimeters. The object is then dropped to the bottom of the container and weighed again. The new weight, minus the weight of water and the container, divided by the volume, is the specific gravity of the object. After fixation, dehydration, and infiltration with the embedding medium, the process should be repeated to determine shrinkage or swelling. The surrounding embedding medium must be blotted off before weighing. To immerse paraffin-embedded specimens, the organ must be pushed into the water with a rod or wire. The two sets of volume measurements thus obtained will provide a correction factor for future stereological results for small organs. For larger organs, which must be studied by subsampling methods or in large slabs, direct volume determination is useful for calculating fixation shrinkage or swelling. For many organs, particularly the kidney, and to a smaller extent the brain, the organ swells when immersed

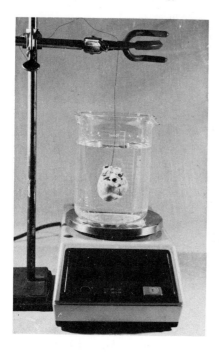

Fig. 3.3. Volume determination by the immersion and weighing method (Archimedes principle). A heart is immersed in water by a wire and the weight read on a balance. [Reproduced with permission from Elias, H. and Hyde, D.M.: Elementary stereology. Am. J. Anat. *159:* 411–446 (1980).]

in an aqueous fixative due to the hypertonicity of the organ. Subsequently, shrinkage occurs during dehydration and paraffin embedding. In the kidney the initial swelling is almost precisely reversed by the subsequent shrinkage. Therefore, an investigator should compare the final volume with the initial, fresh volume, disregarding intermediate volume changes.

Determination of Volume by Serial Planimetry

After an organ is sectioned serially, a *complete* series of slabs of equal thickness can be measured planimetrically. If the area of slab i equals A_i, then the sum of all of these areas multiplied by their common thickness, t, equals the volume of the solid:

$$\boxed{V = t \sum_{i=1}^{n} A_i} \tag{3.1}$$

If the slabs are of slightly different thickness, t in formula (3.1) can become $\bar{t}$, the average thickness.

Direct planimetry is most easily performed by superimposing a grid of test points in square arrangement over the slab. Any specific test point *hits* or does *not hit* the structure to be measured (fig. 3.4); each test point represents the area of the little square formed by four test points. Hits of each position of the grid are counted, transposing the grid frequently. By repeating this procedure several times, the point count will give the area to any desired degree of precision. Doubtful points, such as those that fall on the periphery (like the little square in fig. 3.4), are counted as one-half.

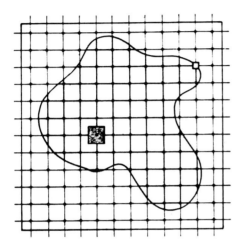

Fig. 3.4. Planimetry by counting point "hits." [Reproduced with permission from Elias, H.; Pauly, J.E., and Burns, E.R.: Histology and human microanatomy; 4th ed. (Wiley, N.Y. 1978).]

Volume Ratios

The determination of the fraction one tissue component occupies in the total volume of a specimen is the easiest of all stereological operations. It is based on the principle of *Delesse* (1848), which states that the areas of profiles of several tissue components are related as the volumes occupied in space by these components, always assuming random distribution and random orientation of components. Using the point-count method of planimetry, a grid of lines is superimposed over the sections and the grid intersections that hit the component of interest are counted. The equation

$$P_P = V_V \quad (3.2)$$

first devised by *Glagolev* (1934), means: the number of points that hit profiles of the component, divided by the total number of test points, equals the volume fraction of that component in the entire specimen (for example, the number of grid points, in figure 3.5a, divided by 100 equals the total number of test points of the grid in this specific position). To illustrate, figures 3.5a, b, and c show three "sections" of prostate glands at different ages. Grid points hitting profiles of stroma, epithelium, lumen, and corpora amylacea give information on the volume ratios of these components in the organ as a

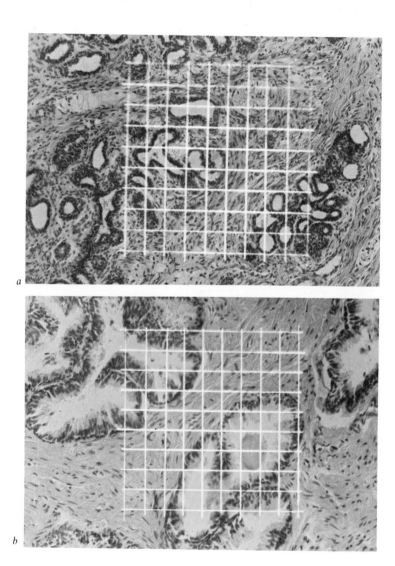

Stereological Measurements of Isotropic Structures

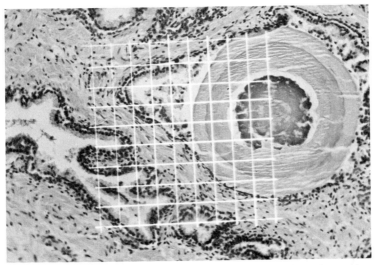

Fig. 3.5. The prostate at three different ages with 100-point grid superimposed to determine the volume fractions of tissue components. Human prostate glands at various ages: *a*, 12 years; *b*, 16 years; and *c*, 45 years.

three-dimensional solid. When a trichrome stain is used, the stroma can be subdivided into muscle and connective tissue.

Volume ratios can also be determined by the *intercept** method. Basically, this means that lines of known length are superimposed on the profiles of the feature, in this case glomeruli (fig. 3.6). Those portions of the lines that fall on a profile (black in fig. 3.6) of the feature to be evaluated are called intercepts. The combined length of all intercepts, divided by the total length of all the test lines used (L_L), is equal to the volume ratio (V_V):

$$L_L = V_V \tag{3.3}$$

The intercept method, introduced in 1898 by *Rosival*, is convenient for very small details, such as the space within mitochondrial cristae or the cisternae of the endoplasmic reticulum, or the volume fraction of interalveolar septa—in short, for volume fraction determination of relatively thin, flat structures whose profiles would rarely be hit by points.

---

* The term *intercept* should not be confused with *intersection*, a term used to determine surface area per volume.

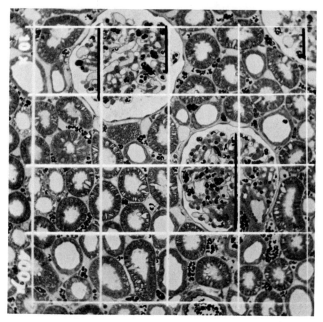

Fig. 3.6. Determination of volume fraction of tissue components by point hits P_P (crossings of white lines) and by intercept measurements L_L (black line segments).

When volume ratios are to be determined by automatic scanners, the intercept method is most convenient. If the specimen is composed of two phases of strikingly different colors (for example, green for connective tissue and red for parenchyma, as in a Goldner stain), a beam of light coupled with a spectrometer can be made to meander over the specimen while the duration of illumination of phases 1 and 2 is recorded. The meandering motion is best performed by an electrically driven mechanical stage, while the light beam remains stationary.

The principle of equivalence of area measurements of *Delesse* (1848) and of intercept measurements, introduced by *Rosival* (1898), later reduced to a point count with volumes by *Glagolev* (1934) seems never to have been proved mathematically, but has been arrived at by various authors, using the commonsense method. The equation

$$P_P = L_L = A_A = V_V$$

is nevertheless mathematically correct. It is of the nature of an axiom and can be justified by using the words of *Euler*: "Mathematics is the science of the self-evident."

One of the most lucid explanations of this principle was given by *Arthur Holmes* (1921). *Holmes* as well as his predecessors *Delesse* (1848) and *Rosival* (1898) and his successor *Glagolev* (1934), were all geologists. Their objects of observation were rocks composed of mixtures of various minerals, each having specific, distinguishable optical properties. Holmes' explanation follows (p. 310 ff):

"Volume-percentages can be estimated with a more or less close approximation to accuracy by means of measurements on a plane rock-surface (such as a thin section)—(a) of the areas of the constituents; (b) of the linear intercepts along a series of lines; provided that the rock is uniform in the distribution of its minerals and homogeneous in texture, and provided that the total area of the surface is sufficiently large, compared with the grain-size of the rock to give a fair representation of the average composition. A rock of this kind has the same appearance on any given surface in whatever direction it is cut . . . .*

. . . (a) the ratio of the sum of the areas of any given mineral to the total area of the measured rock-surface is approximately equal to the volume percentage of that mineral in the rock. Along any line of adequate length drawn on a plane surface.— (b) the ratio of the sum of intercepts of any given mineral to the total length measured across the rock-surface is approximately equal to the volume-percentage of that mineral in the rock. Instead of making measurements on a single line, which would have to be very long and would, therefore, require an extensive surface to contain it, a series of lines at suitable distances apart, or two series of lines at right angles, may be measured up on a much smaller surface.

The validity of the above principles is easily established for uniform (we would say isotropic) rocks. Consider a cube of a uniform rock, sufficiently large to be a representative specimen. Let the cube be cut into square plates. Then, if all the plates are placed side by side in the same plane, the bulk composition is the same as that of the cube, and areas are proportional to volumes. If now the composite plate is cut into narrow prisms of square cross section, the length of each mineral along the prism is proportional to its area, and, therefore, to its volume. Then, if all the prisms are placed, end to end in the same straight line, the bulk composition is the same as that of the cube, and linear intercepts are proportional to volumes. Strict equality can be achieved by the above process only when the plates or prisms are infinitely thin."

The next simplification of the principle was accomplished by *Glagolev*, 13 years later (1934). He showed that a lattice of points superimposed on the material was often more practical to use than the linear intercept method. *Chalkley* (1943) reinvented the point-hit method 9 years after *Glagolev*.

---

* Nowadays this property is more frequently described as isotropy.

Before the introduction of point counting and intercept measurement, a seemingly time-consuming method was often used: the images of sections were projected on thick paper and cut out with scissors, then the pieces that had been cut out were weighed. Professor Friedrich Wassermann used to take the tracings home before a big party and distribute drawings and scissors to his guests to speed up the otherwise boring work.

A study of bronchial gland volumetric density compared the accuracy and mean measurement time for the methods of planimetry, point count, and paper cut-out weighing (*Bedrossian et al.* 1971). Surprisingly, the accuracy was equivalent for all three methods, but the paper cut-out weighing method required *half* the time of the other methods. Therefore, depending on the complexity of the structure to be quantified, simpler shapes may be more easily estimated by the paper cut-out weighing method.

Gray/white volume ratio is a problem that has come into focus since 1972 (*Gazzaniga*) when the bilaterality of the brain was recognized. Stereology offers a great new tool to penetrate deeper into the quantitative differences of both hemispheres, a potential source of information to knowledge yet untapped. Superimposition of counting patterns on complete series of brain slabs of this kind could become of considerable interest.

Core-to-Shell Volume Ratio

Pannese et al. (1972, 1975) developed an interesting method to estimate the ratio of the volume of unipolar neuronal cell bodies in spinal ganglia to the volume of the satellite cell sheaths, and found significant correlations between these ratios and the taxonomic position of the animal. In the following paragraphs an analogous method is presented. It was developed from Pannese's procedure, using a different, but structurally similar cell group as a paradigm. Results are not shown; only shown is how it could be done by anyone interested in the subject. In fact, it is hoped that someone will pick up this suggestion and come up with worthwhile results.

Our paradigm is growth of the primary ovarian follicle having a primary ovocyte (oocyte) as its core and a layer of follicle cells as a shell (fig. 3.7a–c). While the entire complex grows, the volume of the shell increases more rapidly than that of the core, as is obvious from mere visual inspection of these three pictures.

The primary ovocyte as well as the entire follicle approximate spheres in shape. But since they are slightly oval, *Pannese* introduced the following approximation: the area of the profile of the object is measured planimetrically, assuming that the section is near equatorial; and, the follicle as

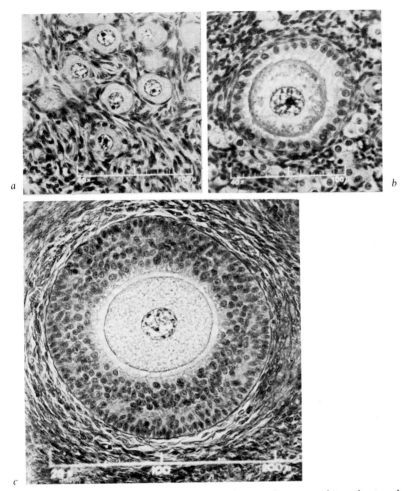

Fig. 3.7. Ovarian follicles in different stages of maturation are used to estimate volume ratios at successive developmental stages of the shell/core. Core = oocyte; shell = community of follicle cells.

well as the ovocyte are assumed to be nearly spherical. *If* the formation were spherical, then its radius would be $r = \sqrt{A/\pi}$. The sphere's volume would be $V = \frac{4}{3}\pi r^3$. The volume of the entire follicle would be called V_f (V_{follicle}) and that of the ovocyte alone V_c (V_{core}). The volume of the shell is $V_s = V_f - V_c$. To obtain the best possible results from single sections when serial sections are not available, only such follicles are selected in which the nucleolus of the ovocyte has been hit. Such sections are nearly equatorial. These profiles vary but little from truly equatorial sections, since 31.2% of all possible sections of a sphere have radii r so that $r = R \pm 5\%$, a mistake that is quite tolerable. This astonishingly high percentage of sectional circles almost as large in radius as the cut sphere was determined as follows: Draw a sector of the unit circle (fig. 3.8). Mark along the base of this quarter circle, which has the radius 1, the radius of the smallest circle to be tolerated. It measures 0.95 and equals the cosine of the angle θ formed by the base of the quadrant and a line connecting the center with the periphery of the smallest acceptable circle. Arc cos $0.95° = 18.2°$, the sine of which is 0.312. Therefore, the probability that a sectional circle cut through a sphere of radius $R = 1$ has a radius $r \geq 0.95$ is 31.2%. It is, therefore, quite probable, since the nucleolus is never far away from the center of an almost spherical ovocyte, that the radius r of a sectional circle showing the nucleolus equals $R \pm 5\%$.

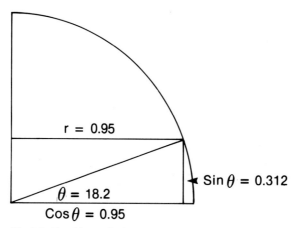

Fig. 3.8. Graphic method to determine the percentage of sections through a sphere whose diameter does not deviate more than 5% from that of the sphere itself.

All of this can be done without any knowledge of trigonometry simply by drawing figure 3.8 on graph paper and measuring cos θ. Thus, it can safely be assumed that the radius of the section based on the above mentioned planimetric method is very close, though always a tiny bit smaller than the radius of the whole cell. Since the same mistake is made on the entire follicle, the calculated volume ratio V_c/V_s (V_c meaning the volume of the core; V_s

meaning the volume of the shell) is very close to reality. We are dealing here not with a precise ratio determination but with a very close estimate.

Surface Area Per Volume

An organ may contain a great number of parts of the same kind (such as follicles, alveoli, or capillaries), and it may be necessary to estimate the total surface area through which diffusion can take place.

However, before describing the technique for the determination of surface area per unit volume, let us present a derivation of the standard formula,

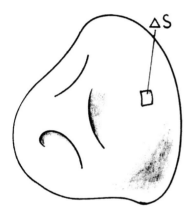

Fig. 3.9. Definition of a surface area of an isotropic body, the surface exhibiting no preferential orientation: [Reproduced with permission from Elias, H.; Pauly, J.E., and Burns, E.R: Histology and human microanatomy; 4th ed. (Wiley, N.Y. 1978).]

which was originally worked out by *Smith and Guttman* (1953) using a different method. This particular derivation was presented for the first time in 1971 by *Elias, Hennig,* and *Schwartz,* and the following step was taken by the senior author in 1 hour and 20 minutes.

Imagine a surface curved in space that does not show any preferential orientation (fig. 3.9). Each of its elements ΔS can be placed, without loss of orientation, on the surface of a sphere, since each specific orientation is as frequent as any other one. In such a case of isotropy the surface can be cut apart and all the elements of that surface rearranged on a sphere of radius r (fig. 3.10). This sphere may be contained in a cube of edge length ℓ. Divide the upper facet of this cube into ℓ^2 little squares, each one unit wide and one unit long. Imagine that parallel light rays fall on the sphere from above so that the sphere casts a shadow on the bottom of the cube. The area of the shadow is πr^2.

Then draw a vertical line through the center of each little square on the top facet of the cube. Each line has the length ℓ. A total of ℓ^2 lines pass

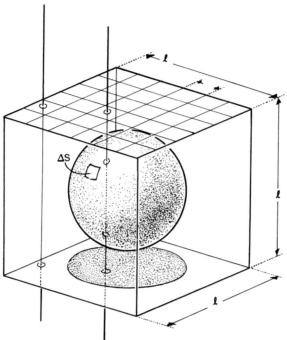

Fig. 3.10. Derivation of S_V formula. [Reproduced with permission from Elias, H.; Hennig, A., and Schwartz, D.E.: Stereology. Application to biomedical research. Physiol. Rev. *51*: 155–200 (1971).]

through the centers of all of the little squares on the top. Therefore, the combined length of all these test lines is $L = \ell^3$. The surface of the sphere is $S = 4\pi r^2$. Each test line intersects the surface of the sphere twice but its shadow only once. Thus, lines of total length $L = \ell^3$ have achieved $P = 2\pi r^2$ intersections with the surface of the sphere. This surface is contained in the volume ℓ^3. Since $S_V = 4\pi r^2/\ell^3$ and $2\pi r^2 = P$, and the total length of intersecting lines is $L = \ell^3$, the latter two values can be substituted to obtain

$$S_V = 2P_L \tag{3.4}$$

In this simple calculation, r and π have been canceled out.

An example of the determination of the surface density of hepatic sinusoids is shown in figure 3.11. An array of lines of known length is superimposed on a section through a liver. The intersections of these lines with the sinusoid boundary are counted and used as P in formula (3.4). Five of these points are emphasized by black circles in figure 3.11. In this manner, the sinusoidal filtration area per unit volume of liver can be determined.

Stereological Measurements of Isotropic Structures 39

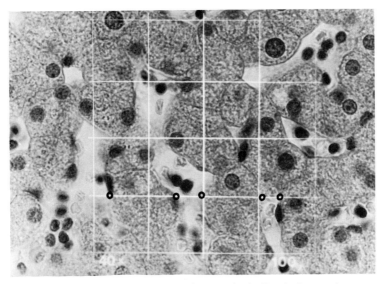

Fig. 3.11. Determination of a sinusoid surface area in the liver by intersection count.

Figure 3.12 shows another example: here we are determining the area of the glomerular basement membrane per glomerular volume.

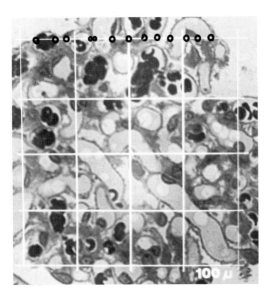

Fig. 3.12. Determination of the glomerular filtration area by intersection count.

Another method for determining surface area per unit volume uses a curvimeter to measure the length L of a trace per unit area on sections. Once L is known, S_V can be calculated as follows:

$$\boxed{S_V = \frac{4L}{\pi A}} \qquad (3.5)$$

Although this method is less efficient and usually less accurate than the method outlined above, it is mathematically just as valid. This latter method is often used when a computer tracing tablet is employed. The difficulty of trace-length measurement lies in the fact that the human hand is not steady enough.

The *absolute* surface area of a solid or of an individual particle can be determined if a *complete* series of sections or slabs exists and if the average thickness t of the slices or the slabs is known (fig. 3.13). Absolute surface

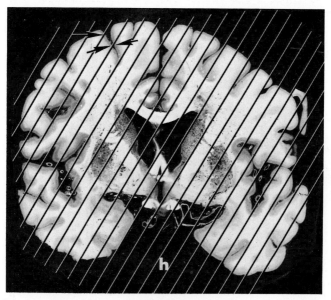

Fig. 3.13. Determination of the *absolute* value of the cerebrocortical surface area by intersection counts (arrows). In this case, the series of slabs of known thickness must be complete, the test lines must be equidistant; but their length does not need to be known.

determination is accomplished by superimposing an array of parallel, equidistant lines over the sections. It is not necessary, in this case, to know the length of the test lines, but the distance of their separation, h must be known. *All* intersection points on *every* slab must be accumulated. In such cases,

$$\boxed{S = 2Pht} \tag{3.6}$$

where P is the number of intersection points of these lines with the traces of the surface of the object. As an example of the determination of the absolute surface area for an object of complicated shape we present the cerebral cortex (fig. 3.13) and use formula (3.6). Intersection points P of the cerebral surface with test lines of constant distance h are counted. The thickness of the slabs is t. *The series must be complete to obtain accurate results*. This method was employed to examine 50 specimens from 22 mammalian species, including 21 humans, and interesting taxonomic correlations were found by *Elias et al.* (1969) and *Elias and Schwartz* (1969 and 1971).

In this connection it is important to note that surface determinations depend on the magnification used. This was pointed out by *Keller et al.* (1976). The higher the magnification, the greater P_L seems to become, because more details become visible. *Weibel* (1979) has compared this with the problem of determining the length of the coast of England.

Length Per Unit Volume

It is often interesting to know the total length of a tube or of fibers or of the edges of space-filling polyhedra such as fat cells, of which only profiles can be seen in sections. Other examples are the average length of proximal convoluted tubules, the combined length of the seminiferous tubules, or the combined length of capillaries in a bulky organ. A problem of physiologic interest may require such data for its solution.

In former times, length per unit volume was determined by maceration and teasing out of long objects, stretching them out on a glass plate, and measuring them with a yardstick. This method presents technical obstacles that are often insurmountable. Before the advent of stereology, combined maceration and teasing was the only method available for length determination. Admirable work in that field has been done by *Oliver* (1939 and 1968) as well as by *Sperber* (1944). These investigators teased out whole nephrons from acid treated kidneys and measured the length of each. Using the method of maceration and teasing, the lengths of *individual* tubules are obtained, while stereology deals with averages only. *Oliver* (1968) determined the length of many individual nephrons in normal human kidneys as well as in various kidney diseases. *Sperber* (1944) applied the same method to kidneys from 40 mammalian species ranging from *Monotremata* to *Cetacea* and

Primates. In exceptional cases, if a linear structure is present in the singular, such as the ductus epididymidis, the entire intestine, the total length of brain convolutions, etc., stereology can be applied even to individual structures using formula 3.6. However, using the technique of Alexander the Great, the stereologist can cut the Gordian knot instead of untying it. By counting the number of intersected elements per unit area, the length of the rope is easily obtained.

Derivation of $L_V = 2P/A$

The following derivation of the formula for length per volume is taken from *Hennig* (1963b). It is a very elementary derivation. Imagine a curved line of length L twisted within a standard volume V. It is bent in every possible direction of space without any preferential orientation. This curve can, mentally, be cut into n very short pieces ΔL, so short that, for practical purposes, they are straight. Let us move all these fragments, each retaining its spatial orientation, so that their center points fall together. The end points of these line fragments will come to lie on the surface of a little sphere of radius r. Now, the specimen is cut into slices of thickness $t = r = \Delta L/2$ (fig. 3.14). Every surface element of the sphere ΔS = constant contains an

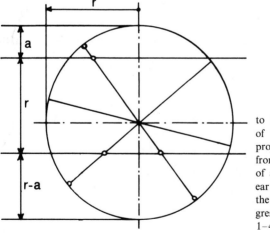

Fig. 3.14. A diagram to illustrate the derivation of the L_V formula. [Reproduced with permission from Hennig, A.: Length of a three-dimensional linear tract, Proceedings of the First International Congress on Stereology; pp. 1–44 (Vienna 1963).]

equal number of end points. On whichever level within the specimen the little sphere is located, it will be intercepted twice by cutting planes, since $r = t$. Following our conventional symbolism, the number of intersection points is designated by the letter P. Therefore, the number of intersection points per unit area A with the line segments equals the number of end points of the line elements ($P = 2n$) (fig. 3.10). The combined length of the line segments is L. Thus:

$$\boxed{L_V = \frac{2P}{A}} \quad (3.7)$$

P is the number of intersection points of the curve with the cutting planes. The factor 2 had to be introduced because in the spherical model each line segment had two end points. In reality, however, the segments are joined end to end so that one of the two end points is no longer an end point; and the number of intersections is only half the number of segment end points.

The above derivation assumes that the space-curve be a line of thickness zero, and that the cutting planes be equally thin. However, *Hennig* (1963b) has shown that the formula applies also to thick tubes and to thick slices.

Consider as an example, the length of the ductus epididymidis (fig. 3.15). We superimpose a test square of known area A over the image and count the number of profiles of tubules within the area. Profiles which project over the

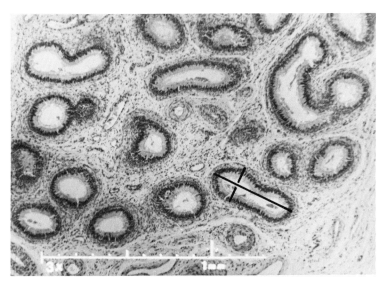

Fig. 3.15. Human epididymidis for determination of the length of the duct by profile counts per test area (use scale) and for shape determination by axial ratio measurement (see cross at lower right).

left and upper side of the square are considered "in;" those intersected by the right and lower sides are considered "out." The number of profiles equals P (Q in Weibel's system) and the total test area used equals A. The length of ductus epididymidis per unit volume is:

$$\boxed{L_V = 2P_A} \qquad (3.8)$$

However, since the thickness and relative amount of interstitial tissue varies from head to tail, sections of the epididymidis should be taken at several levels.

The *absolute* length of a system of linear structures can be determined by counting all the profiles on every slab of a complete series of slabs. This length is given by

$$\boxed{L = 2Pt} \qquad (3.9)$$

A practical example of the application of this method is shown in figure 3.16. This picture illustrates how the total length of the gyri of the cerebral cortex, superficial and hidden, can be determined by counting all the cortical convexities (darts) on a complete series of slabs of average thickness t. Simply count these convexities (P) on one side of each slab and use formula 3.9 to calculate the total length of all the gyri.

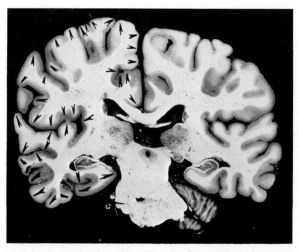

Fig. 3.16. Determination of total length of cerebral gyri by a count of convexities (darts). This count must be made on a complete set of slabs of known thickness. [Reproduced with permission from Elias, H., and Hyde, D.M.: Elementary stereology. Am. J. Anat. *159:* 411–446 (1980).]

IV. Determination of Shape

Introduction

The question of shape, which was the original incentive for the foundation of stereology (*Elias* 1977), is often not appreciated by scientists who assume that shape is "merely" a qualitative attribute. On the contrary, shape must first be known before size and numerical density can be determined. Shape is also important because the function of many parts of living organisms is closely related to their shapes. For example, a muscle must be oblong to exercise its function of shortening. A glandular alveolus must be wide and round to permit the accumulation of secretion. A lung alveolus must have a similar shape to accommodate a specific amount of air. A duct must be a long cylinder to be able to conduct liquids to their destination and to permit the action of its epithelium to modify its content. In essence, shape is the primary property of a structure because knowledge of the shape of a structure is a prerequisite in its measurement. In this section the quantitative basis of shape determination is discussed.

Points, Lines, and Surfaces

According to the principle of dimensional reduction, a point (granule) will not be cut by a mathematical plane, except in extremely rare cases. However, in histology, where slices rather than true sections are involved, granules will be found within a slice, between its two cutting planes. A granule will be in focus at oil immersion with the condenser up and the diaphragm open for a depth of 0.2 μm only. An object can be identified as a granule when it can be brought into and out of focus within the "section."

A line intercepted by a plane yields a point. Fibers, fibrils, and filaments have the quality of lines. A slice of finite thickness might contain a short segment of fiber. When cut perpendicularly, this fiber segment will appear in the microscope as a point that remains in focus throughout the thickness of the slice. A fiber segment of length L inclined toward the cutting plane will appear as a rod whose apparent length varies with the angle of inclination θ.

Its projection length L', as observed in the microscope (*with diaphragm narrow*), is

$$L' = t \cot \theta \qquad (4.1)$$

(fig. 4.1). With the condenser high and the diaphragm open, the position of the inclined fiber segment will appear to shift sideways when focusing up

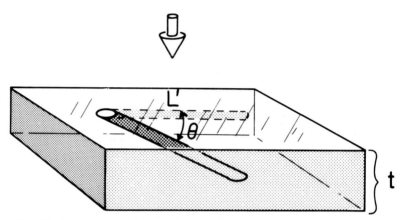

Fig. 4.1. Apparent length L' as a fiber segment included in a slice and inclined against the cutting plane would exhibit. [Reproduced with permission from Elias, H. and Hyde, D.M.: Elementary stereology. Am. J. Anat. *159:* 411–446 (1980).]

and down. In electron microscopy, on the other hand, a segment of a filament or microtubule will remain sharp throughout its intercepted length, because of the great depth of focus of that instrument. Formula 4.1 could be used in these instances if the angle of inclination is known. The length of microtubules and filaments in high magnification TEM is treated in Chapter 9.

Applying the rules of probability, it can be shown that if a mass of fibers, filaments, or microtubules randomly running through space is cut with a microtome at thickness t, approximately 50% of the fiber segments included in the slice will appear shorter than t, and about 50% will appear longer. Very long lines will be rare; i.e., about 1% of all fiber segments will appear longer than $7t$ and only 0.5% longer than $10t$. On superficial inspection, the long lines appear more numerous than they are because each occupies more area than several dots together. Therefore, actual counts must be made within specified areas to identify fibers, filaments, and microtubules correctly. Because of the considerable thickness of an "ultrathin section" compared with the width of the field, and because of the great depth of focus, the rules

Determination of Shape

just spelled out must be weighed more carefully and more rigorously in electron microscopy than in light microscopy.

Since fibers are often curved, they may wind within the slice, thus raising the number of longer segments above 50%. When the number of lines longer than t exceeds 60%, it is probable, in light microscopy, that among the suspected fibers there are, in reality, some bands.

A two-dimensional surface intercepted by a plane yields a line called a "trace" of the surface. Membranes often have the geometric quality of surfaces. Hence, if one finds numerous lines in a slide but few dots and commas, it is probable that these lines are traces of membranes (*Elias and Spanier* 1953).

Solids of Various Shapes

Fortunately, most organs are constructed of components of similar three-dimensional shape. Homogeneous or isotropic construction is a physiologic advantage because only similar parts can act in unison. Such organs lend themselves well to geometrico-statistical analysis because of their uniform architecture. (Anisotropic organs will be discussed in Chapter 8.) An exception to this general rule is the mitochondrion, which is highly variable in shape, i.e., pleomorphic. Determination of mitochondrial shape is not possible by current stereological methods, but requires either serial ultrathin sectioning with reconstruction or semithin sectioning under high-voltage electron microscopy and stereoscopic treatment using parallax. *Hoffmann and Avers* (1973) have shown, by reconstruction from serial sections, that in spite of the presence of numerous mitochondrial profiles in thin sections, yeast cells possess a single, basket-shaped mitochondrion.

Most other components of living systems approximate spheres, ellipsoids, disks, cylinders, prisms, or sheets. These basic shapes lend themselves easily to stereological identification, measurement, and quantification. For example, certain nuclei often have indentations and lobes, but these are so compactly arranged that the overall shape is usually simple (fig. 4.2). Nuclei of polymorphonuclear leukocytes are an obvious exception.

The shape of a profile of a solid on sectioning depends on the three-dimensional shape of the solid and on the angle, and sometimes on the level of cutting. Compare a cell nucleus with an equestrian statue and consider how different sections through one equestrian statue could look. Fortunately, rarely are structures of such complexity dealt with in histology. A notable exception would be a renal podocyte. When the configuration of a solid gets close to a geometrically defined body, for example an ellipsoid, only the angle, not the level of cutting, influences the shape of a profile. A solid has

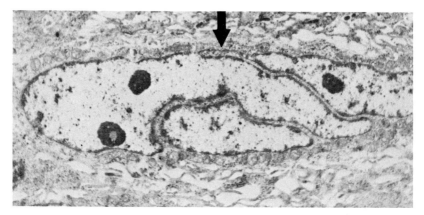

Fig. 4.2. A greatly lobated and indented nucleus of a colon adenoma. The overall shape when viewed by light microscopy would be very simple. [Reproduced with permission from Elias, H. and Hyde, D.M.: Elementary stereology. Am. J. Anat. *159:* 411–446 (1980).]

three dimensions: length, width, and height. Its sectional profile has only two dimensions: length and width. The quotient length/width is called its axial ratio Q. If many three-dimensional objects of equal shape, randomly distributed in space, are cut by a plane, the axial ratios of their profiles will show a characteristic distribution from which the common shape of all these particles can be determined (fig. 4.3).

Spheres

One hundred percent of sections through spheres will have an axial ratio of $Q = 1$ (i.e., they are all circles). Thus, if only circles are found in a slice, they have resulted from cutting numerous spheres.

Circular Cylinders

Circular profiles in slices may also be profiles of parallel circular cylinders cut transversely, such as is seen by looking at the open end of a pack of cigarettes. In histology a transverse cut through a peripheral nerve exhibits this aspect; so does a transverse cut through a medullary ray of the kidney. If the plane of cutting is oblique to the longitudinal direction of these parallel circular cylinders, all profiles will be ellipses of equal shape. Their common axial ratio will be equal to the cosecant of the angle the cutting plane forms with the longitudinal direction of the cylinders:

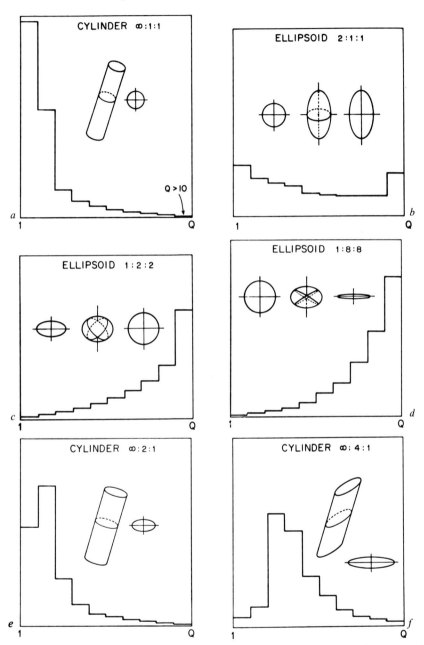

Fig. 4.3. Distribution of axial ratios of profiles of variously shaped bodies. [Reproduced with permission from Elias, H. and Hyde, D.M.: Elementary stereology. Am. J. Anat. *159:* 411–446 (1980).]

$$Q = \operatorname{cosec} \theta \qquad (4.2)$$

An oblique cut through the renal medulla, is an example.

When circular cylinders are randomly distributed and oriented in space (as the portions of the ductus epididymidis, fig. 3.15), 75% of their profiles will be "short" ellipses, i.e., their axial ratios Q will lie between 1 and 2 ($1 \leq Q \leq 2$), and 25% will be more oblong. In histologic "sections," however, the long ellipses will appear just slightly more numerous than they would in cutting planes of thickness zero. In fact, their absolute length will be augmented by $t \cot \theta$. The distribution of axial ratios in sections through randomly cut straight circular cylinders is shown in figure 4.3a. Cylindrical tissue components (such as tubules, arteries, or the gut as a whole) are often confined to a restricted space; as a result, they are often twisted, convoluted, and curved. For this reason, the number of very long ellipses will be even lower than the value of 25% predicted for straight cylinders. Actually, if a cylinder in the human organism is straight, like certain large arteries and ducts, it will usually be solitary (that is, unaccompanied by structures of its own kind) except for nerve fibers and nerve fiber bundles. Randomly arranged cylinders in man are always crooked. This decreases the frequencies in Q classes $8 - \infty$, but increases the frequencies in classes $4 - 6$.

Rotatory Ellipsoids

Intermediate in shape between spheres and circular cylinders are prolate rotary ellipsoids, oval bodies with one rotational axis (a) and a transverse diameter. Rotatory ellipsoids may be classified as either prolate or oblate. Prolate rotatory ellipsoids are oblong, egg-shaped bodies, symmetric about the rotational axis. Nuclei in columnar epithelium exhibit this shape. Oblate rotatory ellipsoids are flattened, lens-shaped bodies symmetric about the minor axis, as nuclei in squamous epithelium. When sectioned in any direction, both kinds yield ellipses. The distribution of axial ratios of their "sections" is characteristic for each type of rotatory ellipsoid, as shown in figure 4.3b–d (b for prolate, c and d for oblate ellipsoids). The ellipsoids are assumed to be equal in shape and randomly oriented, or several sectional planes in random directions are used.

Figure 4.4 shows cumulative curves for axial ratio distributions of prolate and oblate rotatory ellipsoids of any shape, with mean caliper diameters entered; the curves were prepared by *Dr. A. Hennig* (*Hennig and Elias* 1963a).

A summary of additional shapes was made by *Elias et al.* (1971). The Q distributions were worked out by *Elias* and others in a number of papers published in the *Zeitschrift fuer wissenschaftliche Mikroskopie*, quoted in *Elias et al.* (1971).

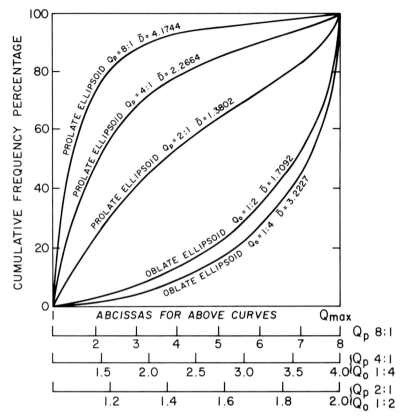

Fig. 4.4. Cumulative distribution curves for axial ratios Q of profiles of ellipsoids of various shapes, prepared by Dr. Hennig. The scale for each form changes with Q max, which is the ratio of largest to smallest extension of a body. [Reproduced with permission from Elias, H. and Hyde, D.M.: Elementary stereology. Am. J. Anat. *159:* 411–446 (1980).]

Triaxial Ellipsoids

Triaxial ellipsoids are oblong, flattened ovoids. As their name suggests, they have three axes: a longest axis (a_{max}), a shortest axis (a_{min}), and an intermediate axis (a_{int}). Many endothelial nuclei exhibit this shape. Sections of ellipsoids with specific degrees of flattening and elongation should show characteristic distributions of axial ratios. To determine such shapes, it remains necessary to produce or search for exact longitudinal and cross sections. *Greeley et al.* (1978) determined these shapes by wax plate reconstruction, from serial, semithin sections, for various nuclei in the lung. *Hennig and Elias* (1963b) tried an empirical approach by cutting artificially-cast ellipsoids of predetermined shape, then measuring and tabulating the axial ratios of the profiles. This was a coarse attempt. Much more work is needed to establish satisfactory guidelines for the shape determination of triaxial ellipsoids from the axial ratios of their sectional profiles.

Elliptical Cylinders

In contrast to circular cylinders whose exact cross sections are circles, those of elliptical cylinders are ellipses (fig. 4.3), so that they approach the shape of bands. The *Taeniae coli* have this shape; so do many veins and venous sinuses, as well as certain *Platyhelminthes*. Distribution of axial ratios for sections of elliptic cylinders of various flatnesses are shown in figure 4.3e and f.

Index of Folding

Another shape parameter is the degree of wrinkling or folding. Figure 4.2 shows a "section" through a nucleus of a colon goblet cell. In the light microscope, this nucleus would appear as an oval. The degree of wrinkling can be determined by superimposing an array of lines over the image. Drawing lines over this electron micrograph would obscure detail, so the reader is asked to perform the following exercise mentally, or with a pencil and tracing paper: Draw a smooth line around the entire profile of this nucleus (it will approximate an oval). Then superimpose an array of equidistant, parallel, straight lines over the image. Their distance and length do not need to be known. Count the number of intersections P of the test lines with the traces of nuclear envelope. Note also that at the point of the arrow, there exists a thin bridge covered on *both* sides by nuclear envelope. Lines crossing this bridge intersect it twice. Then count the number of intersections p with the enveloping smooth line. The quotient

$$F = \frac{P}{p} \tag{4.3}$$

is the *index of folding*. This index assumes great significance in the comparative anatomy of the cerebral cortex, where it is determined on macroscopic slabs of brains (*Elias and Schwartz* 1971). The reader is invited to perform the same exercise on figures 4.5 and 3.13.

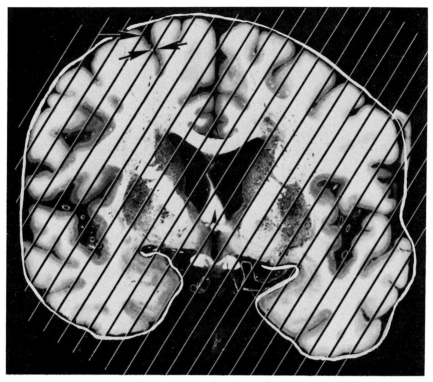

Fig. 4.5. Determining the index of folding F of the cerebral cortex by dividing the number of intersection points P of the test lines with the pia-cortex boundary through the number of intersection points p of the same test lines with the white enveloping line. [Reproduced with permission from Elias, H. and Hyde, D.M.: Elementary stereology. Am. J. Anat. *159:* 411–446 (1980).]

Networks or Plexuses and Branched Cylinders

Up to this point, we have discussed only simple, unbranched cylinders, but in the living organism there exist many cylindrical structures that branch, such as ducts and end arteries. In order to identify branching, it is necessary to find an occasional Y-shaped or a T-shaped figure in a slice. The number N of points of branching per unit volume is proportional to the number n of Y-shaped or T-shaped figures per test area and indirectly proportional to the average internodal length I, the thickness of the slice (if translucent), and the thickness T of the branches. The exact relationship has not yet been worked out. In the case of networks or plexuses, H-shaped figures and loops are also found occasionally.

Laminae

The greatest amount of flatness is possessed by a sheet, plate, or lamina of infinite extension and finite thickness. An epithelium or mucous membrane without villi has these properties. While a basement membrane often appears in the light microscope as a surface without thickness, in electron microscopy it acquires the geometric properties of a lamina of measurable thickness. If such a flat object reaches beyond the boundaries of the microscopic or electron microscopic field, it is said to be of "infinite" extension. All of its "sections" are stripes of "infinite" length, whose width W varies with the angle of sectioning.

Muralia

A muralium that consists of interconnected walls appears in sections as a maze of interconnected stripes without end, in any direction. A classic case of muralium is the liver. In fact, the study of liver structure provided the impetus for the foundation of stereology. Numerical rules for the identification and topologic classification from random sections of muralia have not yet been worked out, but muralia are easily recognized by the presence of numerous loops in sections. Muralia exist in several tissues and organs: "spongy" bone is seldom a system of spicules but more often a muralium osseum. The seminal vesicles are muralia mucosa, as is the wall of the cystic duct (*Elias, Pauly, and Burns* 1978).

Branched Sheets

Intermediate between a muralium and a very flat, oblate lens is a branched lamina of finite extension, such as the vascular lamina of the renal glomerulus. Its sections are very long, branched stripes with an axial ratio distribution intermediate between those for a very flat, oblate lens and for a sheet of finite extension (*Elias* 1957).

Curvature

The mean surface curvature of a tissue element is a characteristic of shape and can be determined stereologically. It assumes importance in problems of coherence of a tissue, for example when the interdigitations of epithelial cells are considered, and in problems of surface increase. To determine the

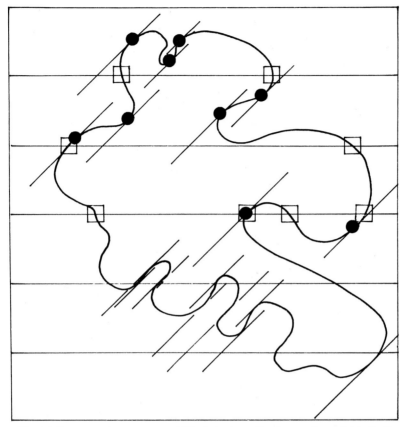

Fig. 4.6. Tangent points are counted (black dots) as a line is moved parallel to itself over the surface of a profile to determine mean surface curvature. The sweep line should be distinguished from intersection points (squares) of a test line. [Reproduced with permission from Elias, H. and Hyde, D.M.: Elementary stereology. Am. J. Anat. *159:* 411–446 (1980).]

mean surface curvature $\bar{K}$, sweep a line remaining parallel to itself over the slice and count the number of tangent points T of the trace with the moving line (black dots in fig. 4.6). Then make an intersection count P/L. Intersection points P are shown as empty squares in figure 4.6.

$$\bar{K} = \frac{\pi T A \text{ net}}{2P_L} = \frac{\pi \cdot T_{net} \cdot L}{2PA} \tag{4.4}$$

gives the mean surface curvature (*DeHoff* 1967). For a definition of the term curvature see Chapter 1.

In his original paper, DeHoff uses the difference between the number of convex tangent points and the number of concave ones and calls it $T_{A\,net}$ ($T_{A\,convex} - T_{A\,concave} = T_{A\,net}$). In the above formula, therefore, the value $T_{A\,net}$ would appear instead of the total number of tangent points T_A.

This account exhausts the analysis of shapes known to play a role in human organs which had been considered stereologically up to 1981. As new objects are observed, hitherto unknown shapes requiring further analysis are likely to be discovered.

V. Numerical Density and Mean Caliper Diameter

Introduction

The chief topic of this chapter is "numerical density"—i.e., the number of particles in a standard volume. Figure 5.1 illustrates this concept. Determination of the number N of particles in a unit volume V from a count of their profiles in a sectional area A has caused great difficulties. The simplest derivation of the basic formula was devised by *Hennig* in a paper by *Elias, Hennig* and *Elias* (1961). In Weibel's and in Underwood's books several derivations are reported. However, they are of such great mathematical sophistication that we have felt obligated to devise a new and simple derivation, even simpler than that by Hennig.

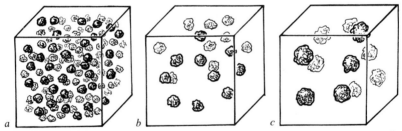

Fig. 5.1. The concept of numerical density. (*a*) Six-month-old baby: 109 glomeruli/mm³; 1,690,000 glomeruli in one kidney. (*b*) Seven-year-old boy—19 glomeruli/mm³; 1,155,000 glomeruli in one kidney. (*c*) Forty-seven-year-old man—11 glomeruli/mm³; 1,309,000 glomeruli in one kidney. Stated densities would be greater for the cortex only. [Reproduced with permission from Elias, H. and Hennig, A.: Stereology of the human renal glomerulus; in Weibel and Elias, Quantitative methods in morphology; pp. 130–166 (Springer-Verlag, N.Y. 1967).]

N particles of average diameter $\bar{D}$ are distributed randomly in a standard unit volume V (fig. 5.2). To make it easy, we begin with a concrete and simple case: the unit volume is a cube of $a = 1$ mm edge length. $V = 1$ mm³ $= a^3$. The particles in this volume are all spheres of equal diameter $D = 100$ μm and they number N. This cube is now cut into 100 slices, 10 μm thick. If

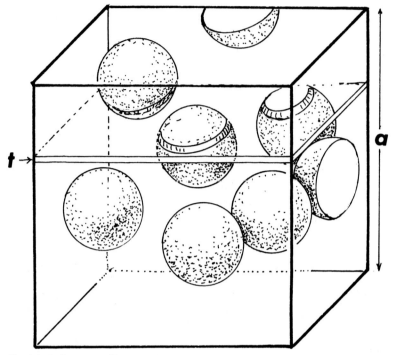

Fig. 5.2. A diagram to illustrate the derivation of N_V formula.

these slices are laid out side by side on a table, they have a total area of $A = 100$ mm^2 = 1 cm^2 and resemble a tile floor. When cut, each sphere yields 10 profiles. Ten times N profiles will be found in an area of 100 mm^2. The total number of profiles in the entire tile floor is $n = 10 \cdot N$. And these circular slices are equally distributed in the area $A = 100$ mm^2. In other words, $n_A = 10N/100$ mm^2. How is N_V related to n_A? Our task is to calculate N_V from n_A, from the section thickness t (which has been 10 μm in this example) and from the common diameter D of the spheres.

In general terms, each sphere has been cut into D/t slices, and the cube has been cut into a/t slices where $a = 1$ mm. When spread out, the slices together measure $A = (a^2 \cdot a)/t = a^3/t$. Since there were N spheres in the cube, $N = ND/t$ profiles were obtained, and these are contained in the area $A = a^3/t$. This can be written symbolically as

$$n_A = \frac{ND \cdot t}{t \cdot a^3} = ND_V \quad \text{or} \quad \boxed{N_V = \frac{n}{A\bar{D}}.} \tag{5.1}$$

In this process the thickness of the slices has been cancelled out, which means that this formula is applicable even to true sections of thickness *zero*, for example the plane of polish in metallurgy. In the latter formula, $\bar{D}$ was substituted for D, out of the awareness that real equality of size rarely, if ever, occurs in living systems. Some spheres may be smaller than D and therefore, yield fewer sections; often they are larger, yielding more sections. Consequently, the average diameter $\bar{D}$ is determined before calculating.

On the other hand, there are objects that deviate in shape from spheres. They may be more oblong, or flat, and are usually located in space at a

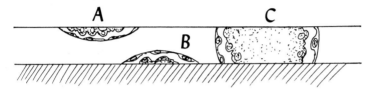

Fig. 5.3. The phenomenon of lost polar caps. Some of them (*A*) may have fallen out of the slice during the cutting and deparaffinization process. Others (*B*) may be just too thin or too small to be recognizable.

slant to the cutting plane. For this reason, for all bodies that are not exactly spherical the average caliper diameter will be used (fig. 1.8). This concept will be discussed in detail below.

Hennig, noticing the dependence of n on t, since no sphere's diameter is an exact multiple of t, has corrected this formula. He added the section thickness to D, because small portions of the spheres (polar caps) will usually project into adjacent "sections." He rewrote the formula thus:

$$N_V = \frac{n}{A(\bar{D} + t)}. \tag{5.2}$$

On the other hand, the polar caps are often unrecognizably thin or fall out of the slice through the mechanical process of cutting or when the paraffin which holds them is dissolved (fig. 5.3). Therefore, he subtracted twice the height h' of these lost polar caps. Twice because it is likely that a corpuscle protrudes on both sides into the adjacent slices. And the height of these lost caps averages $\frac{1}{2}t$. His final highly sophisticated formula became

$$N_V = \frac{n}{A(\bar{D} + t - 2h')}. \tag{5.3}$$

To be precise, one should use formula 5.3 with reasonable estimates of section thickness and the height of lost polar caps.

Since the majority of our sectioning is accomplished using a plastic medium, semithin sections (0.5 to 2.0 μm) for light microscopy and ultrathin sections (50 to 80 nm) for transmission electron microscopy will be discussed but measurements of section thickness using interference microscopy will not be described. Instead, very practical methods are presented, which can be employed in most laboratories.

For semithin sections, a series of sections of varying thicknesses are embedded for any ultramicrotome used. Dense tissues (i.e., cardiac or renal tissues) are used for this purpose. These sections are subsequently resectioned at a precise perpendicular orientation and the section thickness measured in the transmission electron microscope. An excellent discussion of this technique is presented by *Loud et al.* (1978) and by *Ohno* (1980b).

For ultrathin sections *Small's* (1968) method of minimal folds is very practical and precise. In this method, the smallest or minimal fold of a section is photographed. Usually, a faint central line will be visible in this fold. If so, this minimal fold represents a pinching of the section in which the total fold thickness equals twice the section thickness. Even though these two techniques of estimating section thickness are relatively simple, they are also very precise.

Small's method is equally applicable in light microscopy. For example, when a "section" is folded over, as in figure 5.4a, thickness is easily measured simply by focusing on the fold with the oil immersion lens and measuring (fig. 5.4c) the thickness of the section directly. A stereogram of such a sample is shown in figure 5.4a. The arrows in this double picture point to the same spot. More frequent than the fold shown in figure 5.4 are more complex folds such as seen in figure 5.5a. A stereogram of it is shown in figure 5.5c. This fold has a baggy, hollow, triangular portion. "Section" thickness can be measured exactly as in figure 5.4, in the area indicated by arrows. This

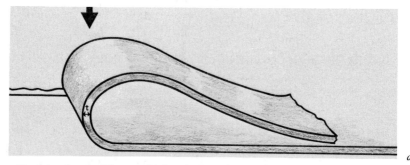

Fig. 5.4. a, b and *c*. The use of defects in "sections" to measure the thickness of a section. In this case the slice is folded over.

Fig. 5.4. b and c. (continued)

conventional triangular fold possesses, at its apex (right side of figure 5.5a), a structure of the same kind as that described by *Small* and which *Weibel* calls a "Small-fold," naming it after the discoverer and at the same time indicating its size. Both open arrows point to the identical spot. These pictures as well as Ohno's observations indicate that the thickness settings of good microtomes are astonishingly accurate. For this reason, a thickness check by the fold method once a year is probably sufficient. We also want to warn the readers against perfectionism: don't discard ugly "sections" because of their folds. As we have demonstrated, such defects can become very useful.

The old method of estimating "section" thickness is provided by the scale of the fine-adjustment screw of a light microscope. Using the oil immersion objective and its full numerical aperture by leaving the diaphragm open and setting the condenser to the highest position, one focuses on the highest visible particle and records the setting of the micrometric focusing screw. Then, one focuses on the lowest visible particle and reads the setting of the fine adjustment again. The difference between both settings estimates

a

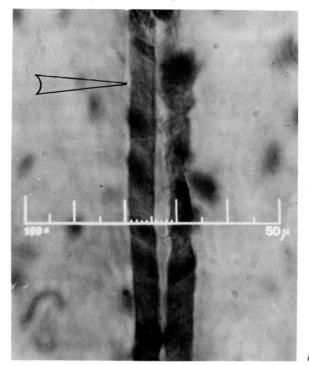

b

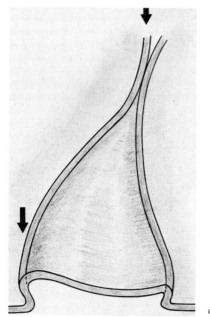

c

Fig. 5.5. (a, b, c) The triangular bag, a very common defect, permits thickness determination (open arrows). At the place of the solid arrow the thickness can be measured as in figure 5.4.

the "section" thickness. This is the only method available for a perfect slide. It is less precise than Small's method.

Estimates of the height of lost polar caps due to low contrast and resolution, and to falling out (fig. 5.3a) can be approached simply by a spherical model. If the tissue is searched for the smallest profile of the structure being measured, it is possible to estimate its distance from the center of the sphere using the Pythagorean theorem and thus determine the height of the lost polar cap. Assuming that R is the true radius of a sphere, r is the radius of its smallest profile, and x is the distance of r from the center of the sphere, then it follows:

$$h^2 = R^2 - r^2 \tag{5.4}$$

since x, the height of the lost polar cap, is

$$x = R - h \tag{5.5}$$

the final formula is

$$x = R - \sqrt{R^2 - r^2} \tag{5.6}$$

In this application $\bar{D}/2$ is assumed to be R, and r is one-half the diameter of the smallest measured profile. This simple application is only valid when all spheres are approximately equal in size. The limits of this estimation are discussed by *Haug* (1967).

In formulas 5.1, 5.2 and 5.3 a bar has been placed above the letter D, because in nature, particles are never uniform in size. The bar validates the formula for numerical density determinations of all kinds of convex particles, the mean caliper diameters of which are known. Since the values t and $2h'$ in the exact formula 5.3 practically annul each other, for most practical applications the modified and very simple equation 5.1 can be used as follows:

$$\boxed{N_V \sim \frac{n}{A\bar{D}}} \tag{5.7}$$

We have stated that the formula is valid for convex particles. This means that it is not valid for certain reentrant particles which, because of their complex shape, might be cut more than once, such as mitochondria (fig. 5.6).

Therefore, numerical density cannot be determined for most mitochondria by stereological means. Reconstructions have been made successfully by *Hoffman and Avers* (1973) of mitochondria in yeast cells using semithin sections and high voltage TEM. They found that a yeast cell possesses one

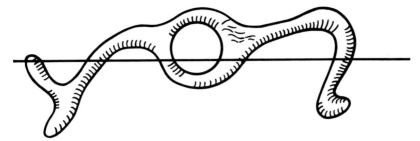

Fig. 5.6. An irregular, reentrant particle (such as a mitochondrion in an embryonic cardiac muscle cell). Numerical density can be determined only if their volumes are known through serial section reconstruction. A profile count will not yield correct results.

single, basket-shaped mitochondrion, which, in "ultrathin sections," will be cut several times.

Size-distribution of Section Profiles

All of the size-distribution methods assume random distribution of particles within a solid or random sectioning of these structures. These methods allow one to determine a size distribution of particles from their sectioned profile size distributions.

Spheres

Let us first consider the case of a group of spheres of equal size. Any section through a sphere is a circle whose radius r depends on the radius R of the sphere itself and on the distance h of the cutting plane from the center of the sphere, so that $r = \sqrt{R^2 - h^2}$ (fig. 5.7). Thus a slice through a mass of spheres will show large and small circles. Even though all profile sizes between $r = R$ and $r = 0$ are possible, the frequency of large profiles is greater than that of small profiles.

To find the common diameter of the spheres we search the "sections" for the largest profile. This profile represents the true radius R of the spheres. As has been shown by *Elias* (1954), if all spheres are of equal size, 13.4% or less of all circles sectioned will have radii r less than $R/2$. Further, if the profiles are divided into four size classes:

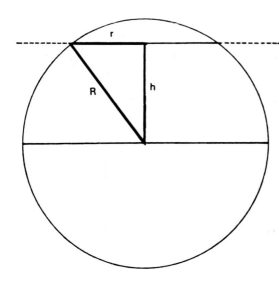

Fig. 5.7. Dependence of the radius of a profile on the radius of the entire sphere and on the distance of the section from the center. [Reproduced with permission from Elias, H.; Pauly, J.E., and Burns, E.R.: Histology and human microanatomy; 4th ed. (Wiley, N.Y. 1978).]

for 3.2% of circles, $0 < r \leq 0.25R$;
for 10.2% of circles, $0.25R < r \leq 0.5R$;
for 20.5% of circles, $0.5R < r \leq 0.75R$; and
for 66.1% of circles, $0.75R < r \leq R$.

This distribution is readily visible in figure 5.8 by measuring the dark vertical lines. If $R = 100$, then the length of each dark vertical line expresses a frequency in percent.

To explain how we arrived at the above data, let us arbitrarily divide the circular profiles into 10 size classes so that for class 1, $0.9R < r_i < R$, etc. We can visualize these 10 classes summarized by

$$R = \sum_{i=1}^{10} r_i \tag{5.8}$$

It is evident from the Pythagorean theorem that

$$h_i = \sqrt{R^2 - r_i^2} \tag{5.9}$$

and

$$\Delta h_i = \sqrt{R^2 - (r_i - \Delta r)^2} - \sqrt{R^2 - r_i^2}. \tag{5.10}$$

Numerical Density and Mean Caliper Diameter 67

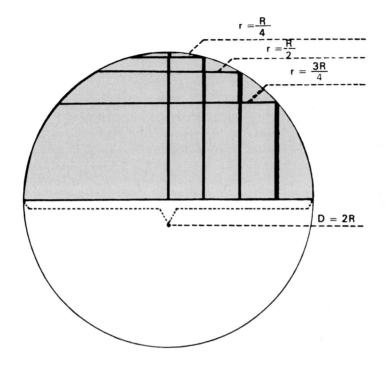

Fig. 5.8. The frequency of the four profile classes when cutting spheres of equal size. [Reproduced with permission from Elias, H.; Pauly, J.E., and Burns, E.R.: Histology and human microanatomy; 4th ed. (Wiley, N.Y. 1978).]

Thus, the frequency of any class of r_i is

$$f(r_i) = \frac{\Delta h_i}{R} = \frac{1}{R} \left[\sqrt{R^2 - (r_i - \Delta r)^2} - \sqrt{R^2 - r_i^2} \right]. \tag{5.11}$$

Hennig's (*Hennig and Elias* 1970) graphic derivation of a frequency distribution of the 20 profile size classes is presented in figure 5.9. But 10 size classes are fully sufficient because more accurate measurements cannot be attained in practical situations.

Since there is an obvious linear relationship between the radius R of a sphere and its mean profile size, we can use the following formula to estimate the true radius from the mean profiles

$$R = \frac{4}{\pi} \bar{r} \tag{5.12}$$

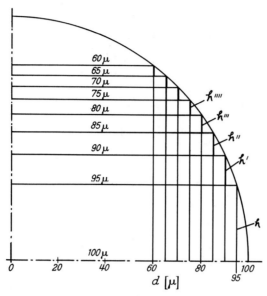

Fig. 5.9. Graphic derivation of a frequency distribution of profiles from cutting spheres of equal size. [Reproduced with permission from Elias, H. and Hennig, A.: Stereology of the human renal glomerulus; in Weibel and Elias, Quantitative methods in morphology, pp. 130–166 (Springer-Verlag, N.Y. 1967).]

This value should correspond to the maximum measured profile, if all spheres are of the same size.

In a graphic representation of profile frequency, some problems are encountered in the lower size classes because of lost polar caps. The easiest way to extrapolate from observed data to recover lost caps is to draw a curve through the mid-points at the top of the last left complete graph bars and complete that curve toward the left so that it passes through the origin. Then draw the formerly empty lost left bars so that the smooth curve passes through the mid-points of their top outlines.

However, in most cases the particles are not of equal size, but show a size distribution so that every particle is neither too small nor too large for proper physiologic function. This is true of such structures as renal glomeruli, thyroid follicles, pancreatic islands, specific cell types, various organelles, and other structures. If more than 13.4% of sectional profiles have diameters d smaller than $D/2$, then from the previous observation, all of the particles in the tissue are not of equal size. The standard curves devised by *Hennig and Elias* (1970) permit an almost instantaneous solution of the problem. The procedure is as follows: diameters of the sectional profiles of particles are classified into 10 size classes, with the upper limits of the classes being $0.1 D_{max}$, $0.2 D_{max}$, $0.3 D_{max}$, ..., $0.9 D_{max}$, and D_{max}. After a sufficient number of profiles (say 1000) have been classified, a histogram (fig. 5.10) is constructed. To each histogram of sectional diameters belongs a smooth curve (black)

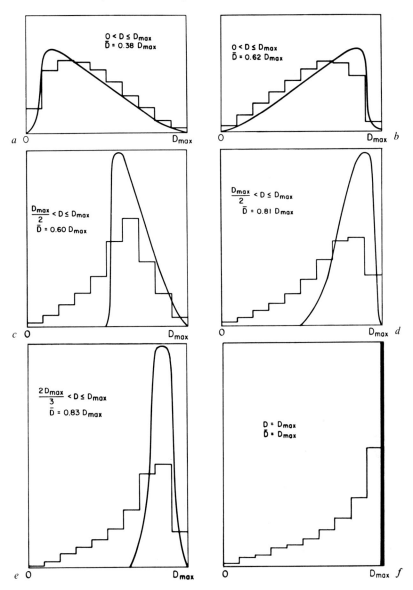

Fig. 5.10. Histograms of frequencies of profile sizes, as they would result from cutting through masses of spheres, showing the size distribution of the smooth, superimposed curves. The size ranges and the average diameter are entered into each field. These graphs permit easy, nonmathematical evaluation of observational material. [Reproduced with permission from Hennig, A. and Elias, H: A rapid method for the visual determination of size and distribution of spheres from the size distribution of their sections. J. Microsc. 93: 101–107 (1970).]

for the size distribution of the particles with an indication of the *average* diameter $\bar{D}$ of all the spheres. This value is indispensable for the calculation of the number of particles per unit volume. Six such combinations are shown in figure 5.10. None of these will ever be perfectly achieved in an actual situation. To arrive at the true size distribution in a specific situation, one can approximate the correct distribution by interpolation. Further size distributions are presented in the original paper by *Hennig and Elias* (1970).

A practical method for determining size classes is as follows: Search several slides, under the microscope, for the largest circular profile. Then couple a particle size classifier to the microscope (Fig. 5.11; *Schwartz and Elias* 1970; *Elias and Botz* 1976). This apparatus, which has never been produced commercially, can easily be built in a laboratory workshop. It consists of an illuminated substage iris diaphragm, the diameter of which is proportional to the arc the diaphragm lever describes between smallest and largest opening. Electrical contact points arranged in a quarter circle represent size classes; each contact point is coupled with an electrical counter. In addition to the 10 classifying contact points, the instrument also has a continuous quarter-circle contact bar. Thus, when the lever is depressed, it activates one classifying counter and the totaling counter. The latter is found immediately near the diaphragm so that one can stop measuring when a standard number of profiles—e.g., 1000—has been reached. The individual size-class counters, however, are connected with the diaphragm by a long cable. The battery of counters cannot be seen by the observer, a feature guarding against human bias. Focus on the largest circular profile and bring the completely open illuminated diaphragm into a position so that it approximates in size the largest profile. This is done with a camera lucida or drawing tube by adjusting the height of the diaphragm from the table and by the zooming device of the drawing tube. The largest sectional circle goes through the equator of the largest spherical particle. Its diameter, to be determined with a micrometer, is D_{max}. To estimate the sizes of very small objects such as nuclei, two mirrors must be inserted between the drawing tube and the illuminated diaphragm. This increases the optical distance and thus reduces the size of the diaphragm's image (*Elias and Botz* 1976; *Elias et al.* 1976).

To begin size classification, superimpose each encountered profile on the image of the illuminated diaphragm, the size of which is adjusted to fit the profile (fig. 5.12). Then depress the lever to activate the counter for the specific size class. After a significant number of particles has been classified, construct a histogram for the frequency of size classes and match it with the best-fitting histogram of figure 5.10. A specific histogram for sectional circles corresponds to a specific distribution of sizes of spheres (drawn as solid lines in figure 5.10). The numbers written into the diagrams give the size range and the average diameter $\bar{D}$ of the spherical particles. This value

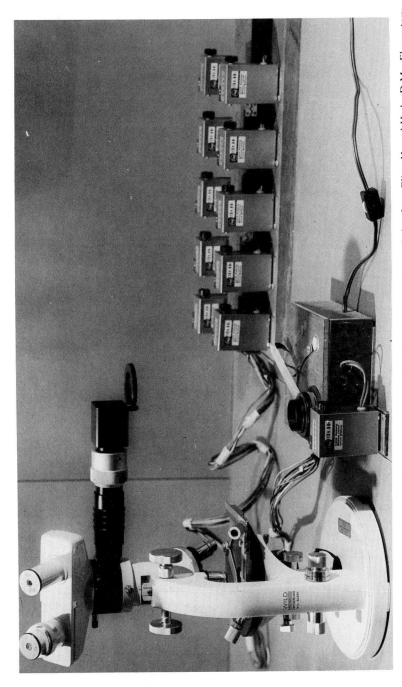

Fig. 5.11. Prototype of the Schwartz-Elias particle size classifier. [Reproduced with permission from Elias, H. and Hyde, D.M.: Elementary stereology. Am. J. Anat. *159:* 411–446 (1980).]

A Guide to Practical Stereology 72

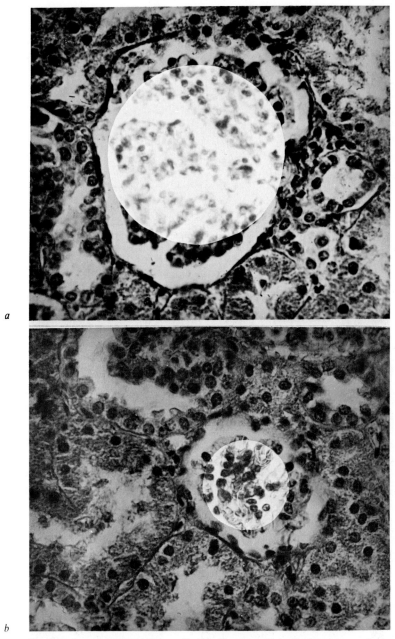

a

b

Fig. 5.12. (*a and b*) Renal glomeruli with illuminated iris diaphragm superimposed for determination of a size distribution.

is particularly important for the determination of the number of particles per unit volume (numerical density). Curves *a* and *b* in figure 5.10 might never be encountered in living systems; curve *a* might approximate sizes of mercury droplets from a broken thermometer or gas bubbles in putrefying cells (*Elias* 1954) and *b* pebbles in a river bed. Curves *c*, *d*, and *e* have a good chance of existing in living systems. Curve *f* represents profiles of spheres of uniform size, like manufactured spheres for ball bearings embedded in a homogeneous plastic; such a condition is rarely encountered in nature.

Profiles of spheroid particles approach the value $D = 0$ toward the poles regardless of the upper size limit of the particles. Therefore, from a mathematical point of view, a curve for sectional distribution always passes through the origin. In practical cases, this is almost never accomplished because the smallest profiles are not recognizable or fall out of the slices. Thus size classes $0-0.1D_{max}$ and $0.1-0.2D_{max}$ are usually empty. (An illustration of this effect, employing renal glomeruli, is given in figure 5.3. Particle A may fall out of the slice, while particle C might be too small or too indistinct for identification.) As a result, algebraic and analytic methods for determining a size distribution of particles can result in negative numbers for size classes of small diameter. Our visual method solves this problem by smoothing out the observational histogram to the left so that the curve passes through the origin, thus compensating for the lost polar sections.

Construction of the distribution of d (diameter of circles) from one of D (diameters of spheres), as presented in figure 5.8, is described briefly. Fortunately, figure 5.9 and formula 5.10 can be used in our explanation. Consider simplistically that in a distribution of spheres in 10 size classes, only those spheres in class 10 can give rise to profile radii in class 10 ($N_{A_{10}}$). From this it follows that a histogram similar to figure 5.10f will be generated using formula 5.11 if also we consider the probability of section profiles in classes 1–9 (N_{A_1} to N_{A_9}) contributed by class 10 spheres. Thus, all profiles counted in class 10 are contributed by spheres of class 10. For $N_{A_9} = N_{A_9}$ ($f_{10} \cdot r_9$), the total profiles in class 9 is multiplied by the frequency or probability of class 10 spheres yielding class 9 radii. Since $N_v = N_A/2R_v$ it follows that:

$$N_{A_9} = N_{V_{10}} \cdot 2R_{10} \left[\frac{1}{R_{10}} \sqrt{R_{10}^2 - (r_9 - \Delta r)^2} - \sqrt{R_{10}^2 - r_9^2} \right]$$
$$= N_{V_{10}} \cdot 2 \left[\sqrt{R_{10}^2 - (r_9 - \Delta r)^2} - \sqrt{R_{10}^2 - r_9^2} \right] \quad (5.13)$$

If this estimation is continued, a histogram similar to figure 5.10f for class 10 spheres will be obtained. This procedure is then repeated for each successive class of spheres until 10 individual histograms have been completed. These

10 histograms are summed to arrive at the final composite histograms presented in figure 5.10a–e. This method can also be accomplished graphically by measuring Δh (fig. 5.9) and using the following formula that conforms to formula 5.13 (*Hennig and Elias* 1970).

$$N_A = N_V \cdot 2 \cdot \Delta h \tag{5.14}$$

The easiest way to understand this process is by inspecting figure 5.9. If the spheres in the mixture have a size distribution from smallest to largest, only the largest spheres can yield the largest profiles. The largest spheres can also yield profiles of smaller sizes. Thus, small profiles on the plane of section can derive from small and large spheres, while large profiles can only be derived from large spheres. This is the reasoning behind all of the methods for determining size distribution, be they algebraic, analytic or graphic.

A variety of other algebraic and analytic methods have been applied to this problem (*Lenz* 1956; *Saltykov* 1958; *Bach* 1959, 1963b, 1963c, 1964, 1965; *Wicksell* 1925). A chapter by *Underwood* in *DeHoff and Rhine's* book *Quantitative Microscopy* in 1968 and his own book in 1970 provide an excellent summary of the methods. Since we have no intention of reinventing the wheel, we will restrict ourselves to a few conceptual comments in reference to these methods. Algebraic and analytic methods based on profile diameter measurements use tables of coefficients in up to 15 size classes, which estimate the probability of obtaining profiles of a certain diameter based on a certain spherical diameter. Other methods utilize measured areas or intercept lengths to estimate the numerical density in a variety of size classes.

Unfortunately, all the above-mentioned methods are designed for spherical or nearly spherical particles. Many particles, however, are anisodiametric. Again take a ruler as an example. Its height differs in different positions. It is lowest when lying flat on the table, highest when standing in a corner. Its average height or mean caliper diameter ($\bar{D}$) is the average of all heights (H) when this body assumes all possible positions in space. The estimation of mean caliper diameter of anisodiametric particles presents difficulties. Yet such estimates are very important to anatomists, since most particles in the human body are nonspherical. For example, anatomists often wish to estimate the mean caliper diameter of cell nuclei. Except in such organs as the liver, the adrenal cortex, the nervous system, and the mucous columnar cells in cervical glands, nuclei seldom approach spherical shape. In columnar epithelium and in smooth and cardiac muscle they approximate the shapes of prolate rotatory ellipsoids (oblong, egg-shaped bodies). Nuclei of mucous salivary glands and superficial cells of nonkeratinizing stratified squamous epithelia often have the approximate shapes of oblate rotatory ellipsoids (lens-shaped bodies), while nuclei of cells in the stratum spinosum and mucous cells in the cervix uteri are not infrequently spherical. Endothelial nuclei are usually long, broad, and thin, approaching the shape of triaxial ellipsoids (oval bodies with three axes).

An approach by *DeHoff and Rhines* (1961) allows one to calculate mean caliper diameter of a *rotatory* ellipsoid if its general shape (prolate or oblate) is known. First, the axial ratio (a/b) of the body is determined graphically from the mean value of the ratio of minor axis (b) to major axis (a) (fig. 5.13).

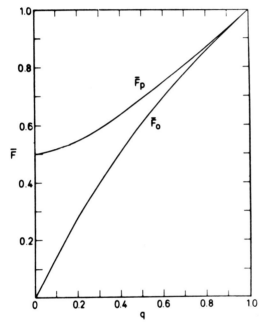

Fig. 5.13. The mean value of the ratio of minor to major axis of cuts through prolate ($\bar{F}_p$) and oblate ($\bar{F}_o$) ellipsoids is plotted as a function of the axial ratio of the generating ellipse. [Reproduced with permission from DeHoff, R.T. and Rhines, F.N.: Determination of number of particles per unit volume from measurements made on random plane sections: the general cylinder and the ellipsoid. Trans. AIME *221*: 975–982 (1961).]

Once the axial ratio is determined, then $\bar{D}$ may be estimated by the following formulas.

For prolate ellipsoids:

$$\bar{D}_{\text{prolate}} = \frac{b}{2}\left[\frac{a}{b} + \frac{b/a \cdot \ln\left(\frac{1+\sqrt{1-(b/a)^2}}{b/a}\right)}{\sqrt{1-(b/a)^2}}\right] \tag{5.15}$$

and for oblate ellipsoids:

$$\bar{D}_{oblate} = \frac{a}{2}\left[\frac{b}{a} + \frac{\sin^{-1}\sqrt{1-(b/a)^2}}{\sqrt{1-(b/a)^2}}\right] \quad (5.16)$$

It is obvious that some size coefficient must be derived since in the above equations absolute values of b and a are called for. *DeHoff and Rhines* define a size factor Z as the reciprocal of the minor axis of elliptical intersections;

$$Z = \frac{1}{b'} \quad (5.17)$$

averaging Z over all orientations, they arrive at

$$\bar{Z} = \frac{\pi}{2b}. \quad (5.18)$$

They further simplify the estimation of $\bar{D}$ and N_V by deriving a shape factor $k(q)$ as a function of the axial ratio of ellipsoids, either prolate ($k_p(q)$) or oblate ($k_o(q)$). This shape coefficient can be estimated graphically from the axial ratio (fig. 5.14). With this coefficient, the following short formulas apply

$$\bar{D}_{prolate} = b \cdot k_p(q) \quad \text{for prolate ellipsoids} \quad (5.19)$$

and

$$\bar{D}_{oblate} = a \cdot k_o(q) \quad \text{for oblate ellipsoids} \quad (5.20)$$

also, an overall formula for N_V can be derived as follows

$$N_{V_{prolate}} = \frac{2N_A \cdot \bar{Z}_p}{\pi \cdot k_p(q)} \quad (5.21)$$

for prolate ellipsoids and

$$N_{V_{oblate}} = \frac{2N_A \cdot \bar{Z}_0}{\pi \cdot q(k_o(q))^2} \quad (5.22)$$

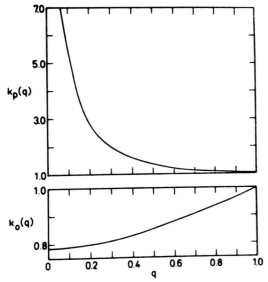

Fig. 5.14. The shape factor is plotted as a function of the axial ratio of the generating ellipse for (a) prolate ellipsoids ($k_p(q)$) and (b) oblate ellipsoids ($k_o(q)$). [Reproduced with permission from DeHoff, R.T. and Rhines, F.N.: Determination of number of particles per unit volume from measurements made on random plane sections: the general cylinder and the ellipsoid. Trans. AIME 221: 975–982 (1961).]

for oblate ellipsoids.

Another set of equations derived independently by *Mack* (1956) 5 years earlier yields identical results:

$$\bar{D}_{\text{prolate}} = \frac{a + b[\sinh^{-1}(a^2/b^2 - 1)^{1/2}]/(a^2/b^2 - 1)^{1/2}}{2} \quad (5.23)$$

$$\bar{D}_{\text{oblate}} = \frac{b + a[\sin^{-1}(1 - b^2/a^2)^{1/2}]/(1 - b^2/a^2)^{1/2}}{2} \quad (5.24)$$

A table of mean caliper diameters estimated for a few selected solids is given below. Both sets of equations by *Mack* (1956) and by *DeHoff and Rhines* (1961) gave identical results:

Prolate rotatory ellipsoids (oblong, egg-like bodies)
  Axes 8:1 $\bar{D} \simeq 4.1744$
       4:1 $\bar{D} \simeq 2.2664$
       2:1 $\bar{D} \simeq 1.3802$
       1:1 $\bar{D} \simeq 1$ (sphere)

Oblate rotatory ellipsoids (lens-shaped bodies)
Axes 1:2 $\bar{D} \simeq 1.7092$
      1:4 $\bar{D} \simeq 3.2227$

DeHoff (1962) also derived an equation that enables one to calculate the number of polydispersed ellipsoidal particles, expressed as the number of ellipsoidal particles in each class interval per unit volume. The particle shape must be constant and a shape factor $k(q)$, must be determined from axial ratios (fig. 5.14). The equation requires that Saltykov's size distribution table (*Saltykov* 1958) be used. This table contains 15 size classes and is reproduced by *DeHoff* (1962) and *Underwood* (1970). The working equation is as follows:

$$(N_V)_j = \frac{1}{k(q) \cdot \Delta} \sum_{i=j}^{n} \alpha(i,j)(N_A)_i \qquad (5.25)$$

where, for any one class size of ellipsoid (j), the numerical density can be determined. Δ is defined as $B_{max} \div n$, where B_{max} is the minor axis of ellipsoids in the largest class size and n the number of size classes; the coefficients $\alpha(i,j)$ are obtained from the Saltykov distribution table; and $(N_A)_i$ is the number of ellipses per profile test area in each size class.

Triaxial ellipsoids differ from rotatory ellipsoids in that they have three axes: a longest axis (a_{max}) or length, a shortest axis (a_{min}) or thickness, and an intermediate axis (a_{int}) or width. Consequently, it is more difficult to determine their mean caliper diameter. One practible method is three-dimensional reconstruction from semithin (0.5 μm) serial sections. Subsequent measurements of wax reconstructions give $\bar{D}$. The following values are approximations for pulmonary cell nuclei using this method (*Greeley et al.* 1978; *Crapo and Greeley* 1978):

Triaxial ellipsoids (thin, broad, and long bodies)
Axes $12:7:2\frac{1}{2}$ $\bar{D} \simeq 8$
      $11:7:4$ $\bar{D} \simeq 8$
      $9:6\frac{1}{2}:5$ $\bar{D} \simeq 7$
      $11:7:4\frac{1}{3}$ $\bar{D} \simeq 8$
      $11:7\frac{1}{2}:4$ $\bar{D} \simeq 8$
      $10:7:5\frac{1}{2}$ $\bar{D} \simeq 8$

DeHoff and Bousquet (1970) presented a method for estimating the size distribution of triaxial ellipsoidal particles from a distribution of linear intercepts using profiles of sectioned structures, provided the particles are all the same shape. This method has been applied to metallurgical structures;

however, its application to biologic structures is absent from the literature. The method can be applied to oriented, partially-oriented, and randomly-oriented structures. For randomly-oriented structures, "sections" are scanned for the smaller axial ratio, $Q_2 = a_{min}/a_{max}$, a number of linear intercepts are measured, and the larger axial ratio, $Q_1 = a_{min}/a_{int}$, can be obtained by evaluating an equation that requires integration of the measured distribution of linear intercepts and graphically reading the corresponding value of Q_1. The number of triaxial ellipsoidal particles in a series of class intervals per unit volume can then be calculated after the shape factor ($\bar{K}_3$) is determined graphically using Q_1 and Q_2.

DeHoff and Rhines (1961) also describe a method for determining the N_V of truncated right circular cylinders. The working equation is

$$N_V = \frac{4 N_{AC} \bar{Z}_c}{\pi^2} \qquad (5.26)$$

where N_{AC} is the total number of straight line segments of cylinder outline (those that intersect at least one end of the cylinder) and Z_c is a size measurement. Z_c is defined as

$$\bar{Z}_c = \frac{\pi}{4 \bar{r}_c} \qquad (5.27)$$

and $\bar{r}_c$ is the mean of the lengths of straight line segments. *DeHoff and Rhines* (1961) also provide formulas for determining the mean radius, length and caliper diameter of cylinders. The mean radius is:

$$\bar{r} = \bar{r}_c = \frac{\pi}{4 \bar{Z}_c} \qquad (5.28)$$

the mean length is:

$$\bar{\ell} = \frac{\pi^2}{4 \bar{Z}_c} \left[\frac{2 N_A}{N_{AC}} - 1 \right] \qquad (5.29)$$

where N_A is different from N_{AC} in that it is the *total* number of intersections with cylinders. Finally the mean caliper diameter is

$$\bar{D}_c = \frac{\pi}{2} \bar{r} + \frac{1}{2} \bar{\ell} \qquad (5.30)$$

Another stereological method for estimating N_V deserves special consideration because of its application to a wide variety of structures when its coefficients of shape and size are estimated by three-dimensional reconstruction techniques. This method of *Weibel and Gomez* (1962) and of *Weibel* (1963) establishes a relationship among N_V, N_A and V_V using a dimensionless shape coefficient, β, and a size distribution coefficient, k, for each shaped structure. The virtue of the method is its use of easy stereological measurements (N_A and V_V) and its ability to evaluate multiple structures on a section at one time.

The working equation is:

$$\boxed{N_V = \frac{k\sqrt{N_A^3}}{\beta\sqrt{V_v}}} \qquad (5.31)$$

The shape coefficient β is related to the mean caliper diameter and the volume of a structure as follows

$$\beta = \sqrt{\frac{\bar{D}^3}{V}} \qquad (5.32)$$

For ellipsoids of revolution and right circular cylinders, *Weibel and Gomez* have provided a graphic method for determining β if the ratio of length to diameter is known (fig. 5.15). Note that a sphere has a β value of 1.382. This can easily be demonstrated using formula 5.32 as follows

$$\beta = \sqrt{\frac{\bar{D}^3}{V}} = \sqrt{\frac{(2R)^3}{\frac{4}{3}\pi R^3}} = \sqrt{\frac{6}{\pi}} = 1.382 \qquad (5.33)$$

The size distribution coefficient is related to mean caliper diameter by the ratio of the cube root of the mean cube of $\bar{D}$ to $\bar{D}$ as follows

$$k = \left[\frac{\left(\frac{1}{n}\sum_{i=1}^{n}\bar{D}_i^3\right)^{1/3}}{\frac{1}{n}\sum_{i=1}^{n}\bar{D}_i}\right]^{3/2} \qquad (5.34)$$

One final method should be mentioned because it is practical and simple. This is the two-section thickness method of estimating N_V originally proposed by *Abercrombie* (1946) and developed by *Ebbesson and Tang* (1965).

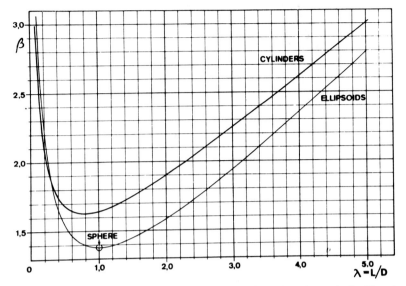

Fig. 5.15. Shape coefficient β for ellipsoids and cylinders of varying ratio λ of length to diameter, $\lambda < 1$ for oblate and $\lambda > 1$ for prolate ellipsoids (Reproduced with permission from Weibel, E.R.: Stereological principles for morphometry in electron microscopic cytology. Int. Rev. Cytol. *26*: 235–302 (1969).

This method as modified and improved recently by *Loud et al.* (1978) is presented. Formula 5.3 can be rewritten as:

$$N_A = N_V t + N_V(\bar{D} - 2h) \tag{5.35}$$

A graph of the numerical profile density (N_A) versus (t) yields a straight line of slope N_V, which is independent of $\bar{D}$ and only slightly influenced by potential unequal loss of polar caps at different section thicknesses (fig. 5.16). The mean volume $(\bar{v})$ of an object can be determined from the following formula:

$$\bar{v} = \frac{V_v}{N_V} \tag{5.36}$$

where V_V is the volumetric density of a structure in the same reference volume as N_V. *Loud et al.* (1978) evaluated this method using cardiac myocytes to estimate $\bar{v}$ and N_V. They counted N_A at five different section thicknesses (1, 2, 3, 4 and 5 μm). Based on the high degree of linearity (correlation

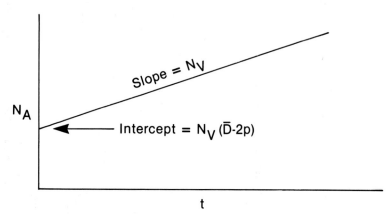

Fig. 5.16. Schematic graph of formula 5.35. [Reproduced by permission from Loud A.V.; Anversa, P; Giacomelli, F., and Wiener, J: Absolute morphometric study of myocardial hypertrophy in experimental hypertension. I. Determination of myocyte size. Lab. Invest. *38:* 586–596 (1978)].

coefficient = 0.944) in the regression of N_A on t, they recommended that determinations of N_A on two sections of known, but different, thicknesses would suffice to estimate N_V.

VI. Thickness of a Lamina and Parallel Cylinders

Thickness of Laminae

A lamina is a sheet, or a wall, or a layer of measurable thickness. Its width and length far exceed its thickness so that in light microscopy or in transmission electron microscopy they extend beyond the boundaries of the field of vision. Its third dimension, thickness, can usually be accommodated within a field of vision.

On cutting, a lamina rarely exhibits its real thickness T, since most sections are oblique, resulting in profiles of the shape of stripes of variable width W depending on the angle of sectioning. The variation is greatest when the lamina is considerably curved. Several methods can be used to estimate its thickness. Let us begin with the assumption that the lamina is everywhere equally thick. Under this assumption, the frequency distribution of width W can be predicted as shown in the cumulative curve (fig. 6.1) or in the histogram (fig. 6.2). In these graphs, multiples of T are plotted on the abscissa

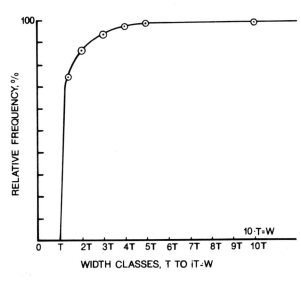

Fig. 6.1. Cumulative curve of width classes that arise from cutting a crooked lamina of thickness T. [Reproduced with permission from Elias, H.: Thickness of a curved lamina. Microscope 28: 67–73 (1980).]

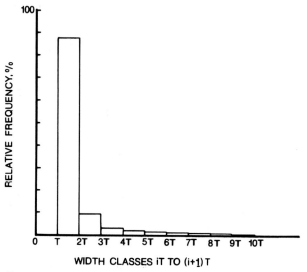

Fig. 6.2. Histogram of width frequencies of sectional stripes arising when a lamina of thickness T is cut. [Reproduced with permission from Elias, H.: Thickness of a curved lamina. *Microscope 28:* 67–73 (1980).]

while the ordinate shows the frequencies in each width class. These classes are $W = T$ to $2T$; $2T$ to $3T$; $3T$ to $4T$, and so on. Again the derivation of the formula and the construction of both graphs were very simple (*Elias* 1980).

Under the assumption of constant thickness, 86.66% of all profiles would have widths between T and $2T$. Only for 13.34% of the places would the sectional stripe be wider than $2T$. This prediction of width distribution is based on the assumption of the divisibility of the total, curved lamina into very many, infinitesimally small fragments, one of which is shown in figure 6.3, like thousands of equally thick table tops floating in space. This is, of course, an unrealistic assumption, because laminae are usually curved within the area of observation. The greater the curvature, the less likely it is that the lamina remains in the plane of cutting for a considerable extent. If a lamina is very crooked, very wide stripes will not occur, and the upper width classes will be empty (fig. 6.4). Both graphs (figs. 6.1 and 6.2) were constructed using the very simple formula:

$$W = \frac{T}{\cos \theta} \qquad (6.1)$$

where θ is the angle which the cutting plane forms with the normal (a line

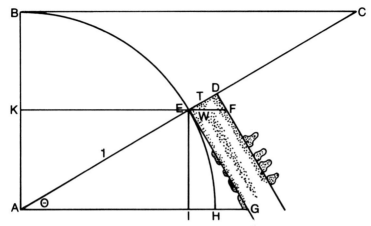

Fig. 6.3. Derivation of sectional width or sections through a lamina of thickness T. [Reproduced with permission from Elias, H.: Thickness of a curved lamina. Microscope *28:* 67–73 (1980).]

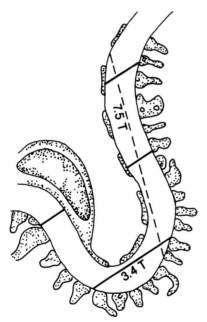

Fig. 6.4. Diagram to show how seldom extreme sectional widths can be achieved. [Reproduced by permission from Elias, H.: Thickness of a curved lamina. Microscope *28:* 67–73 (1980).]

perpendicular) to the lamina. Any person who has read Chapter 1 can derive this formula. When viewing figure 6.3, the triangles EFD and CAB are seen to be similar. Further, the same diagram shows that the frequency of Ws within width classes is proportional to the differences of the sines (line EI in fig. 6.3) between class limits. The procedure can only determine whether the lamina is of constant or variable thickness (see also fig. 6.4).

We have spoken, up to this point, about sheets of rather homogeneous structure, such as the renal glomerular basement membrane. There are, however, other laminae whose thicknesses can be determined from place to place. These are layers which contain circular cylinders or prisms extending throughout the thickness of the layer and directed perpendicularly to it. Two examples are: (a) the mucous membrane of the large intestine and any high columnar or pseudostratified epithelium as long as it contains goblet cells and (b) a third layer of this nature would be the striated border of the intestinal epithelium which contains microvilli of equal height. This latter layer, however, is not yet accessible to the following analysis, because for the magnifications necessary to measure profiles of microvilli, the "ultrathin" sections are too thick.

The thickness of the colonic mucosa and of a columnar epithelium with goblet cells can be determined from place to place. When they are cut normally (perpendicularly), the thickness can be measured directly (fig. 6.5). It is $T = W$. For the large intestine, such places are those where the crypts are cut longitudinally. Even in oblique sections through colonic mucosa of width W, the real thickness T can be determined by finding the axial ratios

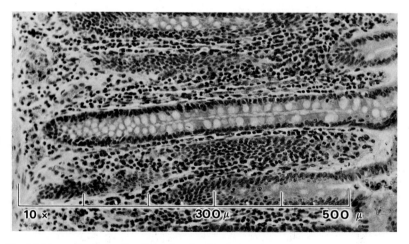

Fig. 6.5. Perpendicular section through colon mucosa.

of the profiles of the crypts (fig. 6.6). Each profile of a crypt has an axial ratio $Q = \text{length/width}$. Their arithmetic mean is $\bar{Q}$. The true thickness of the mucosa can be determined by measuring the distance W from the intestinal lumen to the muscularis mucosae using the formula

$$\boxed{T = W\sqrt{1 - \frac{1}{\bar{Q}^2}}}.$$ (6.2)

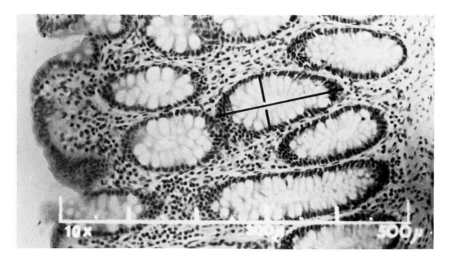

Fig. 6.6. Oblique section through colon mucosa.

Each individual Q is determined most easily by an ocular micrometer that does not need to be calibrated, or on a photographic print with a strip of graph paper, since Q is a dimensionless ratio. Calibration is needed, however, to measure W. Also formula (6.2) was derived by very simple means using only the Pythagorean theorem as shown in figure 6.7.

To determine the thickness of a slab containing an array of truncated cylinders perpendicular to the slab, from oblique sections, we use, again, figure 6.7. As an example for a slab of this kind, we take the mucous membrane of the large intestine, where the cylinders are the glands or crypts. AE in figure 6.7 signifies the surface epithelium; MM (between the two arrows) is the muscularis mucosae. The thickness of the slab is $T = AC$. This slab is cut in the direction AB. Looking at the two similar triangles ABC and $A'B'C'$, it will be noted (according to Pythagoras) that $A'C' = \sqrt{a^2 - D^2}$. The width

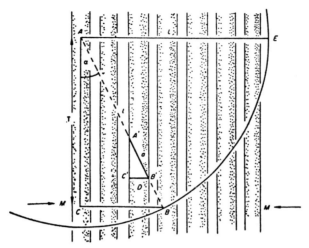

Fig. 6.7. The stereology of parallel cylinders, a diagram applicable to many problems.

of the strip cut out is $AB = \ell$. Because of the similarity of both triangles, $T/\ell = a^2 - D^2/a$; or

$$T = W\sqrt{1 - \frac{1}{Q^2}} \qquad (6.2)$$

figure 6.8 defines Q as the axial ratio of the profile of a crypt. In Chapter 12 this principle is applied to the colonic mucosa and to colonic neoplasms.

Another lamina of physiologic interest is a simple columnar or pseudostratified epithelium. Its thickness can be determined by formula (6.2). However, the "sections" must be $\frac{1}{2}$ μm or thinner. If the epithelium is in the state of water resorption, the intercellular spaces open up because the interdigitations are released as *Kaye et al.*, (1965) have shown (fig. 6.9). Occasionally, a thin, dense layer of ectoplasm may render cell boundaries visible in semithin sections. This is always the case in the urethral glands (crypts of Morgagni, Littre's glands) and in the epithelium of the cervix uteri, even in 5-μm thick paraffin sections. When the cell boundaries are visible, as in the cases just mentioned, T equals W when individual cells are seen to reach from the bottom of the epithelium to the luminal surface (fig. 6.9). Except for urethral glands and the cervix uteri, cell boundaries will usually not be recognizable, because of overlapping of cytoplasm of neighboring cells due to the oblique position of intercellular spaces with respect to the cutting plane. When the slices are thicker, the cytoplasm of the entire epithelium becomes

Thickness of a Lamina and Parallel Cylinders 89

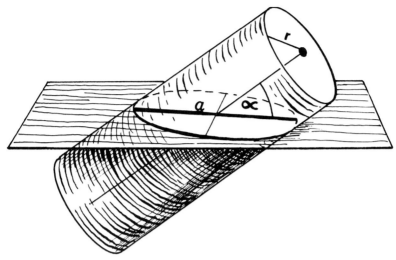

Fig. 6.8. A cylinder of diameter D cut at an oblique angle yields an elliptical profile of axial ratio. [Reproduced with permission from Elias, H.; Pauly, J.E., and Burns, E.R.: Histology and human microanatomy; 4th ed. (Wiley, N.Y. 1978).]

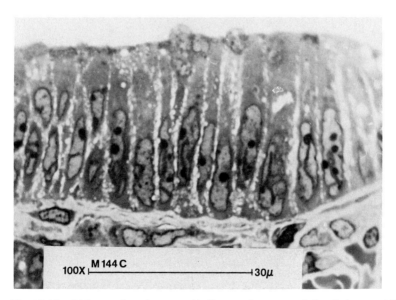

Fig. 6.9. The thickness of a columnar epithelium can be measured directly in a semithin section perpendicular to it, if the intercellular spaces are open as during water absorption (from a colonic adenoma).

an optical continuum. But when goblet cells are present and their basal portions are differentially stained, as in figure 6.10, the epithelial thickness can be calculated by putting the $\bar{Q}$ of cuts through goblet cells into formula (6.2). At favorable spots, where a semithin "section" is normal to the epithelium, one can measure the thickness of such an absorbing epithelium directly, even in the absence of goblet cells, when the intercellular spaces are patent (fig. 6.9), since then one can see the cells reach from the bottom to the free surface of the epithelium. In one such spot, the average height was 31.3 μm (± 3 μm). Even along the short stretch shown in figure 6.9, which is 73 μm wide, there is a variation in height, just a little less than $0.1\bar{T}$. This method was applied to colonic mucosa by *Elias et al.* (1981). At this time we do not know whether any laminae of constant thickness exist in living organisms.

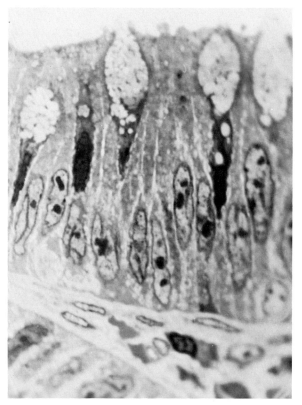

Fig. 6.10. The thickness of a columnar epithelium can be calculated from an oblique section if the axial ratios of goblet cells are averaged.

Weibel (1963) invented an ingenious and very simple method to determine the average thickness of a barrier, based on the fact that the surface area S of a lamina times its average thickness $\bar{T}$ equals its volume V.

$$S \times \bar{T} = V \quad \text{or} \quad \bar{T} = \frac{V}{S} \quad (6.3)$$

One determines, at first, the volume of the lamina within a unit volume V_V by means of a point count $V_V = P_P$. Then one determines its surface area by an intersection count $S_V = 2P/L$. If the lamina is curved and thick, both of its surfaces will be intersected by the same line. When counting both intersection points, we can dismiss the factor 2 and end up with $S_V = P/L$ since each complete intersection has been counted twice. This procedure gives the extent of the lamina along its middle. The average thickness of a curved, thick lamina is

$$\boxed{T = \frac{V_V}{S_V} = \frac{V}{S}}. \quad (6.4)$$

Weibel, however, came up with still a different concept of thickness, considering a lamina as a barrier to diffusion. Such barriers are the air-blood barrier in the lung, the intestinal epithelium, the glomerular basement membrane and others. He considered that diffusion can take place not only perpendicularly, but in any direction. Therefore, the intercepts (L_L) will do (fig. 6.11). In Weibel's system one measures random intercepts, which do not represent thickness but ways of diffusion. Since diffusion is indirectly proportional to the length of the diffusion path, the harmonic mean is the significant quantity. In this system one uses the reciprocal values of the intercepts.

$$\boxed{Z = (1/W_1 + 1/W_2 + 1/W_3 + \cdots 1/W_n) \div n} \quad (6.5)$$

In evaluating the air-blood barrier in the lung, this harmonic mean has proven to be very useful. Intercepts are used only when they run from air to blood, as the two lower lines in figure 6.11. When they run from air to air without intercepting blood (as in the upper line in fig. 6.11) or from blood to blood without intercepting air, they are disregarded.

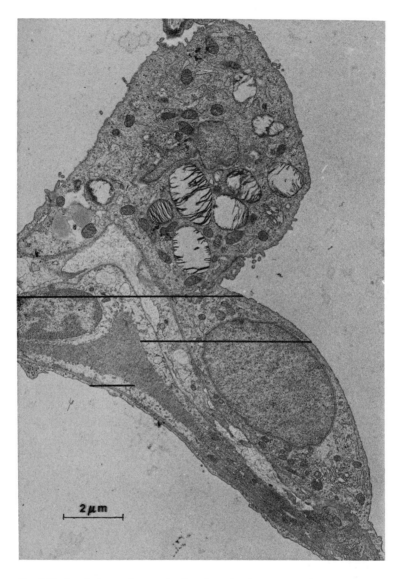

Fig. 6.11. An electron micrograph of an interalveolar septum of a mammal. Random intercepts of the alveolocapillary membrane (black lines) are measured and used in the estimate of harmonic mean thickness, a measurement that relates closely to diffusion capacity where short intercepts are more important than long intercepts. Note the top line is not used because it does not intercept the capillary lumen (micrograph provided through the courtesy of W.S. Tyler).

Parallel Cylinders

Let us assume the presence of many *parallel* right circular cylinders, as for example in a nerve or in a portion of renal medulla. Let these cylinders be cut exactly in a longitudinal direction. At first, we must determine the average diameter $\bar{D}$. This is not always possible since in longitudinal sections one can measure the diameter of a cylinder only when its axis is visible, such as in the case of tubules with a clear-cut, but narrow lumen, or in the case of nerve fibers, with a well-stained axis cylinder. If we find a few axial sections, we can estimate $\bar{D}$.

How to calculate their number per cross-sectional area from a count of their longitudinal sections, has been solved by *Elias* (1976). Their number per cross-sectional area A, i.e., perpendicular to the cutting plane is;

$$\frac{N}{A} = \frac{n}{D \cdot L} \tag{6.6}$$

where L is the total width of microscopic fields measured perpendicularly to the axial direction of the cylinders. Let us return to figure 6.7, and let us note that the cross-sectional area $A = \ell^2$ and that in figure 6.7 $AE = \ell$. In this figure AE represents the area $\ell^2 = AE \cdot \ell$, seen in profile. N cylinders run perpendicularly through the square $AE \cdot \ell$ (cross-sectional area A). If this plane is rotated clockwise to the position AB, $n = N \sin \alpha$ cylinders are cut by this inclined plane, which is now our test area A_T, and we obtain

$$N = \frac{n}{\sin \alpha} = n \cdot \csc \alpha$$

or to be specific

$$N_A = \frac{n \cdot \csc \alpha}{A_T} \tag{6.7}$$

where A is an area transverse to the cylinders, while A_T is the test area on which profiles were counted. Since $Q = \csc \alpha$, we can substitute and arrive at our final formula:

$$\boxed{N_A = \frac{n\bar{Q}}{A_T}} \tag{6.8}$$

VII. Statistical Considerations in Stereology

R. L. Scheaffer

Introduction

The basic stereological relationships between test measurements and global properties of features of interest are discussed in detail in other chapters, but a few of the relationships important to the discussion in this chapter are reviewed here. If the *sample* consists of a two-dimensional test area, the areal fraction (A_A) due to a particular phase is a "good" *estimator* of the volume fraction (V_V) due to this same phase. The estimator obtained from a test area is generally considered to be "good" if the following conditions hold: (a) features that constitute the phase of interest are randomly located and oriented within the volume of mass under study; (b) the test area (sample) is randomly chosen from among all possible test areas that could have been selected; and (c) the test area lies entirely within the volume of mass study. (Two specific properties of "good" estimators will be discussed in the following section.)

Under conditions analogous to (a), (b) and (c) above, it has been observed that L_L is a good estimator of A_A in two dimensions, or V_V in three dimensions. Also, P_P is a good estimator of A_A or V_V, as is $2P_L$ of S_V, $2P_A$ of L_V, $(\pi/2)P_L$ of L_A and $(4/\pi)L_A$ of S_V.

For the statistical considerations to follow, it will be necessary to introduce slightly more notation. Suppose a test area consists of cross sections of features composed of t separate phases ($t \geq 2$). Let

A = total area of the test area

A_i = total area of the features (particles) due to phase i, $i = 1, \ldots, t$.

a_i = area of a single particle from phase i.

L = total length of test line located within the test area.

l_i = length of a single intercept through a particle of phase i.

P = total number of test points located within the test area.

P_i = number of test points falling on phase i features.

Using this notation, $A_i/A = A_A$, the areal proportion for phase i features,

Statistical Considerations in Stereology

$L_i/L = L_L$, the lineal proportional for phase i features, and $P_i/P = P_P$, the proportion of test points falling on phase i features.

Basic Statistical Properties

Suppose a sample is selected from a universe of interest, and a quantity, $\hat{\theta}$, is calculated as an estimator of some quantity, θ, which measures a global property of that universe. Now, $\hat{\theta}$, is a *random variable*, since its numerical value generally will vary from sample to sample. In other words, repeated samples from the same universe will lead to a *distribution* of values for $\hat{\theta}$.

If $\hat{\theta}$ is to be a "good" estimator of θ, one would anticipate that some of the possible values of $\hat{\theta}$ would be less than the true θ and some would be greater, but θ should be close to the "center" of the distribution of $\hat{\theta}$ values. An estimator, $\hat{\theta}$, is said to be an *unbiased* estimator of θ if the *mean* value of the distribution for $\hat{\theta}$ is equal to θ. The notation $E(\hat{\theta})$ will be used to denote mean or expected value, and an estimator $\hat{\theta}$ is then unbiased for θ if $E(\hat{\theta}) = \theta$.

In addition to the "center" of the distribution of possible $\hat{\theta}$ values, one must consider the spread, or variation, in these values. Figure 7.1 shows

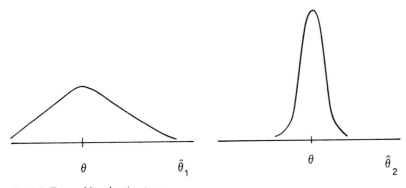

Fig. 7.1. Two unbiased estimators.

possible relative frequency distributions for two unbiased estimators, $\hat{\theta}_1$ and $\hat{\theta}_2$. A little reflection on the part of the reader should convince him that $\hat{\theta}_2$ is the better estimator, all other things being equal, since its values have less variation. Practically speaking, most of the possible values of $\hat{\theta}_2$ are close to θ, but many of the possible $\hat{\theta}_1$ values are quite far from θ. To put this another way, the probability that a $\hat{\theta}_2$ value will be within a small distance of θ is greater than the probability that a $\hat{\theta}_1$ value will be within that same small distance of θ.

Variation in a distribution of measurements is assessed by a quantity called the *variance*, given by:

$$\text{Var}(\hat{\theta}) = E(\hat{\theta} - \theta)^2$$

if $E(\hat{\theta}) = \theta$. Note that this can also be written:

$$E(\hat{\theta} - \theta)^2 = E(\hat{\theta})^2 - [E(\hat{\theta})]^2$$

and is sometimes denoted by $\sigma_{\hat{\theta}}^2$.

To summarize, one generally looks for an estimator, $\hat{\theta}$, which is unbiased and has small variance. One cannot usually find an estimator with arbitrarily small variance, but the variance usually decreases as the sample size increases. Thus, sample size considerations are very important when planning a study.

More specific comments on estimation will now be made with regard to a random sample of measurements, $y_1, \ldots, y_m$. The term *random sample* implies that the y_i's are independent of one another and all have the same distributional properties. In stereology problems, y_i could be a_i, the area of a particle ℓ_i, the length of an intercept A_i, the area of phase i features in a test area, or any similar measurement. Statistical work very often involves the sample mean,

$$\bar{y} = \frac{1}{m} \sum_{i=1}^{m} y_i.$$

If $E(y_i) = \mu$ and $\text{Var}(y_i) = \sigma^2$, then:

$$E(\bar{y}) = \mu \tag{7.1}$$

and

$$\text{Var}(\bar{y}) = \sigma^2/m \tag{7.2}$$

These fundamental properties are used frequently in developing the properties of estimators commonly used in stereology. Examples will be given later.

In addition to properties (7.1) and (7.2), the sample mean, $\bar{y}$, will possess a relative frequency distribution which is approximately normal, if samples of size m are drawn repeatedly from the same universe. That is, the relative frequency distribution of $\bar{y}$-values will look something like the picture in figure 7.2. The normal distribution for $\bar{y}$ is an approximation which generally holds only if m is reasonably large, say $m > 20$ in most cases.

Statistical Considerations in Stereology

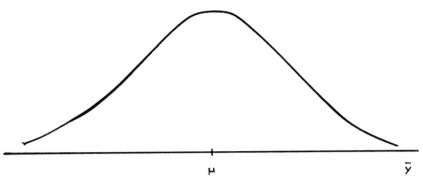

Fig. 7.2. Relative frequency distribution for a sample mean.

Using the normal distribution, one can find for any real number $\alpha (0 < a < 1)$ two numbers, $-Z_{\alpha/2}$ and $+Z_{\alpha/2}$, such that

$$P\left(-Z_{\alpha/2} \le \frac{\bar{y} - \mu}{\sigma/\sqrt{m}} \le Z_{\alpha/2}\right) = 1 - \alpha.$$

This inequality can be reworked into:

$$P\left(\bar{y} - Z_{\alpha/2} \frac{\sigma}{\sqrt{m}} \le \mu \le \bar{y} + Z_{\alpha/2} \frac{\sigma}{\sqrt{m}}\right) = 1 - \alpha$$

The interval

$$\bar{y} \pm Z_{\alpha/2} \frac{\sigma}{\sqrt{m}}$$

is called a *confidence interval* for μ with confidence coefficient $1 - \alpha$. This gives a mechanism for computing intervals which, in repeated usage, should include the unknown μ a high percentage of the time, if α is small. The constant $Z_{\alpha/2}$ is the value cutting off a right-hand tail area of $\alpha/2$ under the standard normal curve, and tables of these values can be found in any introductory statistics book. A common value is $Z_{.025} = 1.96$, or approximately 2.

The confidence interval also gives one a mechanism for calculating a sample size. Suppose that for practical reasons the estimator, $\bar{y}$, should be within d units of μ with high probability. That is, the confidence interval for μ should, in the final analysis, have the form

$$\bar{y} \pm d.$$

Then,

$$Z_{\alpha/2} \frac{\sigma}{\sqrt{m}} = d$$

or

$$m = \frac{(Z_{\alpha/2}\sigma)^2}{d^2} \qquad (7.3)$$

The confidence coefficient $(1 - \alpha)$ will determine $Z_{\alpha/2}$ and σ must be approximated from past experience, theoretical considerations, or a preliminary sample. Even though σ might only be roughly approximated, consideration of m through (7.3) should precede any sampling investigation.

It often occurs, especially in stereological applications, that sampling is accomplished in two or more stages. That is, k primary sampling units, such as test areas, may be selected and then m secondary units, such as test points, are sampled within *each* sampled primary unit. For this sampling design, let

$$y_{ij} = j\text{th measurement on the } i\text{th primary unit.}$$

Then,

$$\bar{y}_i = \frac{1}{m} \sum_{j=1}^{m} y_{ij}$$

is an estimator of a property, μ_i, *on the ith primary unit.*
The average

$$\frac{1}{k} \sum_{i=1}^{k} \mu_i$$

can then be considered as an estimator of a global property, μ, which is in turn estimated by:

$$\frac{1}{k} \sum_{i=1}^{k} \bar{y}_i = \bar{y}$$

In this case, the variance of $\bar{y}$ is composed of two pieces, one assessing the variation among y_{ij}'s within primary unit i and one assessing the variation among the μ_i's. Formally,

$$\text{Var}(\bar{y}) = \frac{1}{k} \text{Var}(\bar{y}_i)$$

where

$$\text{Var}(\bar{y}_i) = \text{Var}\{E(\bar{y}_i \text{ for a specific primary})\}$$
$$+ E\{\text{Var}(\bar{y}_i \text{ for a specific primary unit})\}$$
$$= \text{Var}\{\mu_i\} + E\{\sigma_i^2/m\}.$$

(The term σ_i^2 denotes the theoretical variance associated with an observation selected from the ith primary unit). $\text{Var}(\mu_i)$ will be denoted by σ_B^2 since it reflects a variation *between* primary units. Thus,

$$\text{Var}(\bar{y}) = \frac{1}{k}\left[\sigma_B^2 + \frac{1}{m} E(\sigma_i^2)\right] \tag{7.4}$$

If the σ_i^2's can be assumed equal, the common value will be denoted by σ_w^2 (*within* primary unit variance) and (7.4) becomes:

$$\text{Var}(\bar{y}) = \frac{1}{k}[\sigma_B^2 + \sigma_w^2/m]. \tag{7.5}$$

Note that there are now two sample sizes to consider, k and m, where m depends only on σ_w^2 but k depends on both σ_w^2 and σ_B^2. Careful consideration must be given to both of these choices in terms of cost and in terms of controlling the final variation in $\bar{y}$. If σ_w^2 tends to be large relative to σ_B^2, then most of the experimenters' resources should go toward making m large and k small. If, on the other hand, σ_B^2 is large relative to σ_w^2, then k should be large relative to m. Once m is determined, k can be found through a formula analogous to 7.3, namely:

$$k = \frac{Z_{\alpha/2}^2(\sigma_B^2 + \sigma_w^2/m)}{d^2} \tag{7.6}$$

The ideas inherent in formulas 7.4 and 7.5 can be extended to more than two stages of sampling, but the general case will not be presented here. See *Cochran* (1963) or *Snedecor and Cochran* (1967) for a more detailed discussion.

It remains to show how to estimate the variances used above from information available in a sample or samples. Suppose $y_1, \ldots, y_m$ is a random sample with $E(y_i) = \mu$ and $\text{Var}(y_i) = \sigma^2$. Then a good estimator of σ^2 is provided by

$$s^2 = \frac{1}{m-1} \sum_{i=1}^{m} (y_i - \bar{y})^2$$

$$= \frac{1}{m-1} \left[\sum_{i=1}^{m} y_i^2 - m\bar{y}^2 \right] \tag{7.7}$$

This estimator can be used in the construction of confidence intervals, with the resulting interval being of the form

$$\bar{y} \pm Z_{\alpha/2}(s/\sqrt{m}).$$

In the case of two-stage sampling with y_{ij} denoting the jth observation from primary unit i,

$$\bar{y} = \frac{1}{k} \sum_{i=1}^{k} \bar{y}_i,$$

and

$$\text{Var}(\bar{y}) = \frac{1}{k} [\sigma_B^2 + \sigma_w^2/m].$$

The quantity

$$s_1^2 = \frac{1}{k-1} \sum_{i=1}^{k} (\bar{y}_i - \bar{y})^2$$

provides an estimator of $(\sigma_B^2 + \sigma_w^2/m)$. (The variation among the $\bar{y}_i$'s contains contributions from both σ_B^2 and σ_w^2). The resulting confidence interval estimate then becomes

$$\bar{y} \pm Z_{\alpha/2}(s_1/\sqrt{k})$$

since s_1^2/k estimates $\text{Var}(\bar{y})$.

If it is of interest to estimate σ_w^2 and σ_B^2 separately, σ_w^2 can be estimated by:

$$s_w^2 = \frac{1}{k(m-1)} \sum_{i=1}^{k} \sum_{j=1}^{m} (y_{ij} - \bar{y}_i)^2$$

and then σ_B^2 can be estimated by:

$$s_B^2 = s_1^2 - s_w^2/m.$$

The estimates s_w^2 and s_B^2 can be of help in planning sample sizes m and k for a future study.

Applications to Stereology

Sampling by Test Areas

Suppose that k test areas, each of area A, are randomly selected in accordance with conditions (a) through (c) listed in the Introduction to this chapter. Let $y_j = A_{ij}/A, j = 1, \ldots, k$, where A_{ij} is the total area due to phase i on the jth test area. Then from basic stereological relationships

$$E(y_j) = E(A_{ij}/A) = V_{Vi},$$

the volume fraction for phase i features. Using condition (a), it follows that:

$$E(\bar{y}) = E\left[\frac{1}{k} \sum_{j=1}^{k} y_j\right]$$

$$= E\left[\frac{1}{k} \sum_{j=1}^{k} A_{ij}/A\right]$$

$$= E[\bar{A}_i/A] = V_{Vi}$$

and

$$\text{Var}(\bar{y}) = \frac{1}{k} \text{Var}(A_{ij}/A).$$

To estimate $\text{Var}(A_{ij}/A)$ the measurements $A_{i1}, A_{i2}, \ldots, A_{ik}$ can be used to compute

$$s_{AA_i}^2 = \frac{1}{k-1} \sum_{j=1}^{k} \left(\frac{A_{ij}}{A} - \frac{\bar{A}_i}{A} \right)^2 = \frac{1}{(k-1)A^2} \sum_{j=1}^{k} (A_{ij} - \bar{A}_i)^2$$

and the resulting confidence interval on V_{Vi} becomes

$$\frac{\bar{A}_i}{A} \pm Z_{\alpha/2} \frac{s_{AA_i}}{\sqrt{k}}. \tag{7.8}$$

The interval can be used to estimate V_{Vi} for any one phase i, $i = 1, \ldots, t$.

To estimate more than one V_{Vi} with the same test areas, the confidence intervals should be adjusted somewhat so that the *simultaneous* coverage probabilities for all intervals is approximately $(1 - \alpha)$. There are many methods of making this adjustment, but one simple procedure is the following. If c confidence intervals are to be constructed from the same test areas, with simultaneous confidence coefficient $1 - \alpha$, then use $Z_{\alpha/2c}$ rather than $Z_{\alpha/2}$ for the construction of each interval. For example, if $1 - \alpha = .90$ but $c = 5$ intervals are to be constructed (five different phases are of interest), then use in each interval $Z_{.10/2(5)} = Z_{.01}$ rather than $Z_{.10/2} = Z_{.05}$. The probability that all five intervals then simultaneously include the respective V_{Vi} values should be no less than .90.

In the discussion thus far, $\mathrm{Var}(A_{ij}/A)$ had to be estimated by calculating a sample variance, $s_{AA_i}^2$, from the k observations on A_{ij}. If k is very small, this might not lead to a good estimate of variance and, thus, it would be nice to have a functional form for $\mathrm{Var}(A_{ij}/A)$ as a function of global properties of the specimen under study. Also, some knowledge of the form of $\mathrm{Var}(A_{ij}/A)$ is necessary for sample size considerations, using formula 7.3. Theoretical work by *Hilliard and Cahn* (1961), *Scheaffer* (1969), and others suggest the following formula for the case in which the phase i features show up on the test area as distinct particles with areas a_i:

$$\mathrm{Var}(A_{ij}/A) = \frac{1}{A} V_{Vi}(1 - V_{Vi}) \left[\frac{E(a_i^2)}{E(a_i)} \right] \tag{7.9}$$

Formula 7.9 is equivalent to the one given by *Hilliard and Cahn* (1961) multiplied by $(1 - V_{Vi})$. *Hilliard and Cahn* (1961) were intending V_{Vi} to be small ($1 - V_{Vi}$ to be close to unity) but formula 7.9 works well in many cases with V_{Vi} not necessarily close to zero.

Note that if all phase i particles tend to have equal areas, then $E(a_i^2)/E(a_i) \doteq a_i$, A/a_i is approximately n, the number of phase i particles seen in the test area, and

Statistical Considerations in Stereology

$$\operatorname{Var}(A_{ij}/A) = \frac{1}{n} V_{Vi}(1 - V_{Vi}),$$

which is a binomial-type variance.

In choosing the sample size, k, for the number of test areas, formula 7.9 can be substituted for σ^2 in formula 7.3. Note that k will attain its maximum value when $V_{Vi} = .5$.

The expression $E(a_i^2)/E(a_i)$ in formula 7.9 can be written as:

$$\frac{\sigma_{a_i}^2 + [E(a_i)]^2}{E(a_i)}$$

If a_{ij} denotes the area of the jth particle of phase i, then σ_{ai}^2 can be estimated by

$$s_{ai}^2 = \frac{1}{n-1} \sum_{j=1}^{n} (a_{ij} - \bar{a}_i)^2,$$

where n is the number of phase i particles measured, and $E(a_i)$ can be estimated by $\bar{a}_i = (1/n) \sum_{j=1}^{n} a_{ij}$.

Sampling by Test Lines

Suppose, now, that m random test lines, each of length L, are located within a specimen in accordance with conditions (a) through (c) cited in the Introduction to this chapter. Let $y_i = L_{ij}/L$ where L_{ij} is the total intercept length due to phase i on test line j, $j = 1, \ldots, m$. One can show from basic stereological relationships that

$$E(\bar{y}) = E\left[\frac{1}{m} \sum_{j=1}^{m} y_j\right]$$

$$= E\left[\frac{1}{m} \sum_{j=1}^{m} L_{ij}/L\right]$$

$$= E(\bar{L}_i/L) = V_{Vi}$$

and

$$\operatorname{Var}(\bar{y}) = \frac{1}{m} \operatorname{Var}(L_{ij}/L).$$

$\operatorname{Var}(L_{ij}/L)$ can be estimated by:

$$s_{LLi}^2 = \frac{1}{m-1} \sum_{j=1}^{m} \left(\frac{L_{ij}}{L} - \frac{\bar{L}_i}{L} \right)^2 = \frac{1}{(m-1)L^2} \sum_{j=1}^{m} (L_{ij} - \bar{L}_i)^2$$

and the resulting confidence interval for V_{Vi} is given by

$$\frac{\bar{L}_i}{L} \pm Z_{\alpha/2} \frac{s_{LLi}}{\sqrt{m}} \tag{7.10}$$

For multiple confidence intervals constructed from the same set of test lines, see the comments under *Sampling by Test Areas*.

Formula 7.10 is applicable only if the m test lines are located randomly within the three-dimensional specimen of interest. If, as is often the case, k test areas are first selected and then m test lines, each of length L, are sampled from each of the k test areas, formula 7.4 or 7.5 is appropriate for $\text{Var}(\bar{y})$.

Let $L_{ijs/L}$ denote the proportion of the jth test line within test area s that intercepts phase i features, $s = 1, \ldots, k, j = 1, \ldots, m$. Then

$$\frac{\bar{L}_{is}}{L} = \frac{1}{m} \sum_{j=1}^{m} \frac{L_{ijs}}{L}$$

estimates A_{is}/A, the areal proportion due to phase i on test area s. Also,

$$\frac{\bar{L}_i}{L} = \frac{1}{k} \sum_{s=1}^{k} \frac{\bar{L}_{is}}{L}$$

estimates V_{Vi}.

Assuming that the variances for L_{ijs}/L are approximately equal for all test areas, formula 7.5 will apply and

$$\text{Var}\left(\frac{\bar{L}_i}{L}\right) = \text{Var}\left[\frac{1}{k} \sum_{s=1}^{k} \bar{L}_{is/L}\right]$$

$$= \frac{1}{k}[\sigma_B^2 + \sigma_w^2/m]$$

where

$$\sigma_B^2 = \text{Var}(A_{is}/A)$$

as discussed earlier, under *Sampling by Test Areas*, and

$$\sigma_w^2 = \text{Var}(L_{ijs}/L)$$

for a fixed test area s.

Using the formula for s_1^2 given in the section on *Basic Statistical Properties*, the estimated variance of $\bar{L}_i/L$ is:

$$s_1^2/k = \frac{1}{k(k-1)} \sum_{s=1}^{k} \left(\frac{\bar{L}_{is}}{L} - \frac{\bar{L}_i}{L}\right)^2$$

$$= \frac{1}{k(k-1)L^2} \sum_{s=1}^{k} (\bar{L}_{is} - \bar{L}_i)^2$$

and the resulting confidence interval for V_{Vi} is:

$$\frac{\bar{L}_i}{L} \pm Z_{\alpha/2}\left(\frac{s_1}{\sqrt{k}}\right) \tag{7.11}$$

As discussed in the section on *Basic Statistical Properties*, σ_w^2 can be estimated by:

$$s_w^2 = \frac{1}{k(m-1)} \sum_{s=1}^{k} \sum_{j=1}^{m} \left(\frac{L_{ijs}}{L} - \frac{\bar{L}_{is}}{L}\right)^2$$

$$= \frac{1}{k(m-1)L^2} \sum_{s=1}^{k} \sum_{j=1}^{m} (L_{ijs} - \bar{L}_{is})^2$$

and σ_B^2 is then estimated by:

$$s_B^2 = s_1^2 - s_w^2/m.$$

These estimates can be used in formula 7.6 to determine sample size.

As in the case of sampling by test areas, it is sometimes helpful to know how $\text{Var}(L_{ij}/L)$ relates to global properties of the test area in which the test line is located. A result in *Lackritz and Scheaffer* (1981), which generalizes a result in *Hilliard and Cahn* (1961), shows that

$$V(L_{ij}/L) = \frac{1}{E(M_i)} A_{Ai}(1 - A_{Ai})\left(\frac{1}{t-1}\right)\left\{\sum_{i=1}^{t} \frac{\sigma_{li}^2}{[E(l_i)]^2} + (t-2)\right\} \tag{7.12}$$

where σ_{li}^2 is the variance of the intercept lengths through phase i, $E(l_i)$ is the mean of the intercept lengths through phase i, and $E(M_i)$ is the mean number of phase i intercepts observed on a test line of length L.

If l_{ij} denotes the length of the jth intercept through phase i, then σ_{li}^2 can be estimated by

$$s_{li}^2 = \frac{1}{n-1} \sum_{i=1}^{n} (l_{ij} - \bar{l}_i)^2$$

where

$$\bar{l}_i = \frac{1}{n} \sum_{j=1}^{n} l_{ij}$$

is the estimator of $E(l_i)$.

For theoretical considerations, formula 7.12 is an approximation to σ_w^2 and formula 7.9 is an approximation to σ_B^2 when two-stage sampling with test lines within test areas is employed. A rough idea of sample size can be obtained by substituting 7.12 and 7.9 into formula 7.6.

Note that the variance of L_{ij}/L is directly proportional to the variance among the l_{ij}'s. If intercept lengths tend to be highly variable, then $\text{Var}(L_{ij}/L)$ will tend to be large and the number of test lines needed for a good estimate of A_{Ai} or V_{Vi} will be large. If σ_{li}^2 is small, as for example in sampling a chess board with lines perpendicular to a side, a good estimate is achieved with relatively few lines.

Sampling by Test Points

Suppose that P test points are randomly located within a test area of total area A. Let P_i denote the number of test points which fall upon phase i features. One can write

$$P_i = \sum_{j=1}^{P} y_j$$

where

$$y_j = \begin{cases} 1 & \text{if } j\text{th point falls on phase } i \\ 0 & \text{otherwise,} \end{cases}$$

and thus:

$$\frac{P_i}{P} = \frac{1}{P} \sum_{j=1}^{P} y_i = \bar{y}.$$

Since $E(y_j) = A_{Ai}$ and $\text{Var}(y_i) = A_{Ai}(1 - A_{Ai})$, it follows from formulas 7.1

and 7.2, that:

$$E\left(\frac{P_i}{P}\right) = A_{Ai}$$

and

$$\text{Var}\left(\frac{P_i}{P}\right) = \frac{1}{P} A_{Ai}(1 - A_{Ai}).$$

With A_{Ai} estimated by (P_i/P), the confidence interval estimate of A_{Ai} becomes:

$$\frac{P_i}{P} \pm Z_{\alpha/2} \sqrt{\frac{\frac{P_i}{P}\left(1 - \frac{P_i}{P}\right)}{P}} \qquad (7.13)$$

(This is the usual "large sample" confidence interval for a proportion based on the binomial distribution).

Generally, P points are sampled on k different test areas, and then averaged to form an estimate of V_{Vi}. For this two-stage case, let P_{is} denote the number of points falling on phase i in test area s. Then V_{Vi} is estimated by

$$\frac{\bar{P}_i}{P} = \frac{1}{k} \sum_{s=1}^{k} \frac{P_{is}}{P}.$$

As in sampling by test lines (see formula 7.11) the variance of $\bar{P}_i/P$ is estimated by:

$$s_1^2/k = \frac{1}{k(k-1)} \sum_{s=1}^{k} \left(\frac{P_{is}}{P} - \frac{\bar{P}_i}{P}\right)^2$$

$$= \frac{1}{k(k-1)P^2} \sum_{s=1}^{k} (P_{is} - \bar{P}_i)^2$$

and the resulting confidence interval estimate is

$$\frac{\bar{P}_i}{P} \pm Z_{\alpha/2} \frac{s_1}{\sqrt{k}} \qquad (7.14)$$

It is, again, helpful to construct an exact formula for the variance of $\bar{P}_i/P$ as a function of global properties of the specimen under study.

From formula 7.4, it can be seen that:

$$\operatorname{Var}(\bar{P}_i/P) = \frac{1}{k}\left[\operatorname{Var}(A_{Ai}) + \frac{1}{P} E(A_{Ai}(1-A_{Ai}))\right]$$

$$= \frac{1}{k}\left[\operatorname{Var}(A_{Ai}) + \frac{1}{P} E(A_{Ai}) - \frac{1}{P} E(A_{Ai}^2)\right]$$

$$= \frac{1}{k}\left[\operatorname{Var}(A_{Ai}) + \frac{1}{P} V_{Vi} - \frac{1}{P}(V(A_{Ai}) + E^2(A_{Ai}))\right]$$

$$= \frac{1}{k}\left[\left(1 - \frac{1}{P}\right)\operatorname{Var}(A_{Ai}) + \frac{1}{P} V_{Vi}(1 - V_{Vi})\right]. \tag{7.15}$$

Either formula 7.9 or a suitable estimate can be used for $\operatorname{Var}(A_{Ai})$ if formula 7.15 is to be used to find values of sample sizes k and P. Note once again that increasing the sample size at the second stage (P, in this case) only reduces the second component in formula 7.15. Thus, resources must be used wisely to balance P and k so that $\operatorname{Var}(\bar{P}_i/P)$ becomes suitably small.

It has become quite common to select the P test points within any one test area as the corners of a regular lattice in two dimensions. In this case, the sample is *systematic* rather than random and

$$\operatorname{Var}(P_i/P) = \frac{1}{P} A_{Ai}(1 - A_{Ai}) + C, \tag{7.16}$$

where C denotes a covariance quantity that could be positive or negative (see *Koop* (1976)). Thus, systematic sampling can be better or worse than random sampling.

To illustrate the extremes of systematic point sampling, suppose a regular lattice of points is to be placed on a chess board with red and black squares. If the lattice is chosen so that exactly one point falls in each square, then the proportion of points falling on black squares is a very good estimate of the areal proportion taken up by black. In this case the C in formula 7.16 will be negative and systematic sampling will be better than random sampling.

If, on the other hand, the lattice is chosen so that all points fall on black squares, the sample fraction of points falling on black is a poor estimate of the areal proportion for black. In this case the C in formula 7.16 is positive and systematic sampling is worse than random sampling.

If a systematic array of points is used on a test area that has randomly sized and oriented features, C should be close to zero and systematic sampling should be comparable to random sampling.

Statistical Considerations in Stereology

In short, when using a systematic array of points, the experimenter must carefully check for regular patterns in the features of interest and use the pattern to his advantage. If no regular pattern seems apparent, the systematic array should be treated as a random sample.

The accuracy of the binomial formula for the variance of P_i/P in the systematic case can be checked by taking repeated systematic samples from the same test area and calculating a sample variance (formula 7.7) for the observed P_i/P values.

If two-stage sampling is used with systematic test points within test areas, formula 7.14 is still valid. $\text{Var}(\bar{P}_i/P)$ is still estimated by s_1^2/k even though the array of points within a test area is not random.

Ratio Estimation

All of the estimators considered in previous sections of this chapter have the property that the *denominators* are constants rather than random variables. This property is a result of the sampling design and the particular global properties being estimated. However, for certain global properties it is impossible to structure the sample so that the appropriate estimator has a

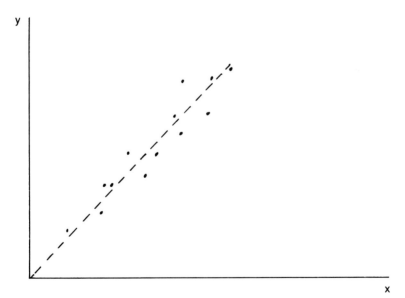

Fig. 7.3. A plot of y_j versus x_j.

constant denominator. For example, suppose three phases of a certain specimen are quite intermingled and all three are likely to show up on any randomly chosen test area. But, it is of interest to estimate the ratio of volume due to phase 1 to volume due to phase 2. Letting A_i denote the area of phase i features in a test area of fixed size A, the volume of interest can be estimated by A_1/A_2. Similarly, $A_1/(A_1 + A_2)$ can be used to estimate the volume proportion of phase 1 among phase 1 and phase 2 features. The properties derived for $A_{Ai} = A_i/A$ do *not* generally hold for estimators of the form A_1/A_2.

In general, let (y_j, x_j) denote a pair of measurements of interest on sample j. A sample could be a test area, a test line, or a set of test points. The measurements could be subareas, intercept lengths, intersection counts or point counts. It is desired to estimate

$$R = \frac{E(y_j)}{E(x_j)}.$$

An estimator of R that works reasonably well in a wide variety of cases is

$$r = \frac{\sum_{j=1}^{m} y_j}{\sum_{j=1}^{m} x_j}, \qquad (7.17)$$

where m is the number of samples. The variance of r can be approximated by

$$\text{Var}(r) = \frac{1}{m\bar{x}^2} \frac{\sum_{j=1}^{m} (y_j - rx_j)^2}{m-1}, \qquad (7.18)$$

where $\bar{x} = (1/m) \sum_{j=1}^{m} x_j$, and a confidence interval for R constructed as

$$r \pm Z_{\alpha/2} \sqrt{\text{Var}(r)}. \qquad (7.19)$$

Generally, r is *not* an unbiased estimator of R. The bias will be small, however, if the plot of (y_j, x_j) values falls along a straight line passing through the origin, as shown in figure 7.3. *Cruz-Orive* (1980) gives an excellent discussion of ratio estimators and their resulting variance estimates. *Miles and Davy* (1976 and 1977), give the general theory for basic stereological estimators and show how to select test areas to reduce the bias inherent in ratio estimators.

A-Weighted Test Areas

In the previous discussions of this chapter it was assumed that the test area (or test line or test points) was randomly located entirely within the boundaries of the mass under study. Edge effects were neglected. (The mass under study could be thought of as "infinitely" larger than the test area.)

In this section the mass under study will be assumed to be relatively small and perhaps irregular in shape. Suppose that a planar region X (see fig. 7.4) contains subregions Y, which constitute the phase of interest. It is desired to estimate $A(Y)/A(X)$, where $A(D)$ denotes the area of the region D, by observing a test area, T. If T is randomly located on X, it is quite likely that $T \cap X$ (the intersection of T with X) will not equal T. That is, some of T is likely to fall outside of X, and $T \cap X$ will have an area which is a random variable.

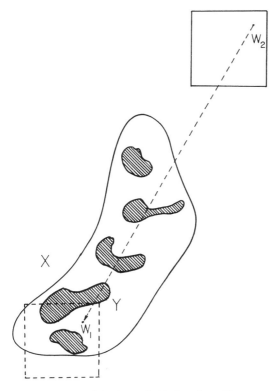

Fig. 7.4. An A-weighted sample T of specimen X.

It should seem natural to use:

$$A(T \cap Y)/A(T \cap X) \tag{7.20}$$

as an estimator of $A(Y)/A(X)$. But the estimator in formula 7.20 is *biased* unless the test area, T, is selected in a particular weighted manner.

An *A-weighted test area* can be selected as follows. Select a point, W_1, at random within the boundary of X. Select a second point, W_2, at random within the boundary of T. T and X are held at fixed but arbitrary orientations. Place the test area T on X so that W_2 is aligned with W_1. The resulting intersection of T with X is an A-weighted test area of X. For an A-weighted test area, formula 7.20 is an *unbiased* estimator of $A(Y)/A(X)$.

If more than one A-weighted test area is used on X, with all test areas of the same size and shape, then let y_j denote $A(T \cap Y)/A(T \cap X)$ for the jth test area $j = 1, \ldots, m$. In this case,

$$\bar{y} = \frac{1}{m} \sum_{j=1}^{m} y_j$$

is an unbiased estimator of $A(Y)/A(X)$ and the variance of y_j can be estimated as in formula 7.7. The resulting confidence interval for $A(Y)/A(X)$ is given by:

$$\bar{y} \pm Z_{\alpha/2} s / \sqrt{m}. \tag{7.21}$$

The theory underlying A-weighted samples is given in *Miles and Davy* (1976) and (1977). The notion of weighted samples has an obvious extension to test lines and systematic arrays of test points.

Stratified Sampling

Suppose a planar region is to be sampled by test areas, each of area A. It is sometimes convenient and often advantageous for purposes of accuracy to divide the planar region under study into nonoverlapping subregions, called strata, and then locate some test areas in each stratum. This is particularly convenient if the region is large with respect to the test areas. In addition, the stratified sampling plan can provide an estimate with smaller variance than that from simple random sampling, if the strata differ considerably with respect to the characteristic being measured. (That is, strata should be selected so that test areas within a stratum give similar measurements, but test areas in different strata give quite different measurements.)

Let a planar region under study be divided into q strata, where T_j denotes the total area of stratum $j, j = 1, \ldots, q$, and $T = \sum_{j=1}^{q} T_j$, denotes the total area of the region. Suppose k_j test areas are located (perhaps randomly) within stratum j. Let $\bar{A}_{ij}$ denote the average of the areas due to phase i on all test areas within stratum j. Then, $\bar{A}_{ij}/A$ estimates the areal proportion due to phase i features in stratum j and

$$\frac{1}{T} \sum_{j=1}^{q} T_j(\bar{A}_{ij}/A) \tag{7.22}$$

estimates the areal proportion due to phase i for the entire region.

If samples are selected independently as one moves from stratum to stratum, then the variance of formula 7.22 is given by

$$\frac{1}{T^2} \sum_{j=1}^{q} T_j^2 \, \text{Var}(\bar{A}_{ij}/A), \tag{7.23}$$

where $\text{Var}(\bar{A}_{ij}/A)$ can be estimated as in *Sampling by Test Areas*.

Of course, if the planar region under study is a random cross section of a three-dimensional structure, then formula 7.22 estimates V_{Vi} also.

Further Statistical Considerations

There are numerous stereological measurements not considered in this account of statistical considerations. If one is interested in intersection counts arising from test lines (used for estimating boundary length and surface area), one cannot use the methods described above for test points. A discussion of some variance approximations in this case is given in *Scheaffer* (1975).

The whole area of particle size distributions is not considered in the foregoing discussion. A good reference on this is *Nicholson* (1970).

The simple two-stage sampling discussed in this chapter can be extended to more complex sampling designs. *Nicholson* (1978) gives a good discussion of the application of analysis of variance to a nested design in stereology. *Snedecor and Cochran* (1967) is an excellent reference for general design and analysis of variance problems.

One final point to keep in mind is that most of the statistical methodology described in this chapter only works for independently selected test areas and independent measurements taken within test areas. When many measurements are taken close together, as in the case of an automatic image analyzer, these measurements become highly dependent, and new statistical techniques must be employed.

VIII. Anisotropic and Nonhomogeneous Tissues and Organs, Extension of Objects

Introduction

Isotropic organs are those of homogeneous structure, so that a random cut through any region of the organ presents the same kind of image. Among such organs are the liver, the lung, the spleen, the large salivary glands, and the prostate. It is impossible, when viewing a slice from an isotropic organ, to identify the region or the direction of cutting. Such organs are ideal for stereological examination.

Anisotropic organs are characterized by preferred orientation of components. Extreme examples of preferred orientation are the parallel arrangement of fibers in muscles and nerves, the cells in the stratum corneum, the pyramidal cells in the cerebral cortex, and the collecting tubules of the renal medulla.

Other organs less strictly organized may show a preferred orientation, such as the convoluted tubules of the renal cortex. Although their course is irregular, these tubules "prefer" a direction perpendicular to the capsule due to their restriction in space between medullary rays.

Estimates of volumetric density by point counting methods are not influenced by anisotropy (*Hilliard* 1962). However, estimates of volumetric density by lineal or areal measurements may depend on the angle of the section plane relative to the primary direction of anisotropy, and the placement of test lines and test areas. Linear intersection estimates of surface density are subject to the same problems.

In Chapter 12 the stereological treatment of a partially oriented tissue—the heart—is illustrated. In cases of partially oriented structures, we prefer to select one tissue block oriented as parallel as possible to the visible orientation and a second block oriented perpendicularly to the first (i.e., longitudinal and transverse sections). In short, we recommend two cutting directions perpendicular to one another. On one set of sections, a single array of test lines and on the other two arrays of lines perpendicular to each other are used, so that we end up with three arrays of test lines which, in space, run in three directions (fig. 8.1). A specific grid for S_V determination,

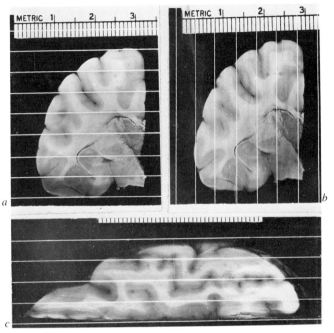

Fig. 8.1. Transverse (*a and b*) and horizontal section through the brain of a coyote (*Canis latrans*). (*a*) Horizontal test lines for intersection count. (*b*) Vertical test lines. (*c*) Longitudinal test lines for intersection count.

often used for both cutting directions, is the curvilinear grid of *Merz* (1967) (fig. 3.1i) which can compensate for anisotropy only, where a preferred orientation manifests itself on one plane of section. It does not free us from the obligation of using a second cutting direction perpendicular to the first. For the Merz grid, the test line length L_T is determined by the following formula:

$$L_T = \frac{P_T \cdot \pi \cdot d}{2} \tag{8.1}$$

where d is the distance between test points P_T (diameter of the semicircles used). An alternative approach is to rotate an orthogonal grid on several fields of a micrograph (*Sitte* 1967; *Mobley* and *Page* 1972). Indeed, it is not infrequent for anisotropy to manifest itself on a single cutting plane. Methods simpler than the curvilinear pattern of Merz exist to compensate for planar anisotropy (*Sitte* 1967; *Mobley and Page* 1972). One of the early compensating patterns was that by *Weibel* (1963), shown in figure 3.1g, which consists of many short line segments forming a 60° angle against each other. A second system used for intersection counts, which the senior author has used frequently in the past, especially for P_L counts on electron micrographs, was an array of equidistant, parallel straight lines of known length. After each count, the pattern was rotated 60°. The parallel lines drawn on a sheet of plexiglass were of unlimited extension. Their cumulative length, no matter which direction the pattern was superimposed on a print, always remained the same within the area of a photographic print so long as the parallel lines extended beyond its edge on all sides. In our opinion our standard, square pattern with 16 fields, 25 crossing points and 10 lines, is much easier to use. The pattern, in a single position, yields two directions perpendicular to each other. By rotating it 45°, we have four directions. Rotating it 30° each time yields six directions. At any rate, a rectilinear pattern is always easier to use than a curvilinear one since the eye follows straight lines without difficulty. Rotating the slide, if you are fortunate enough to have a microscope with a rotating stage, is better than rotating the pattern.

If one has a very anisotropic tissue such as skeletal muscle, the direction of anisotropy can be used as the orientation axis for intersection, counting at the correct or "average" angle. Initially proposed by *Sitte* (1967) and subsequently by *Eisenburg et al.* (1974), the optimum orientation angle (θ) that gives the average orientation between 0° and 90° is 19° (and 71°) from the oriented axis. This mean value of the section angle (θ) satisfies the following equation:

$$\cos\theta + \sin\theta = 4/\pi. \qquad (8.2)$$

For anisotropic structures, *Underwood's* Chapter 3 on oriented structures (1970) is an excellent review. We have used his formulas for partially oriented lines in space to estimate the degree of orientation of a specific system of fasciculated tubules, the coronary capillaries (Chapter 12). The working formula is:

$$\Omega = \frac{(P_A)_\perp - (P_A)_\parallel}{(P_A)_\perp + (P_A)_\parallel}, \qquad (8.3)$$

where Ω represents the percentage of three-dimensional orientation and P_A are profile counts per unit test area in two orthogonal planes ($\perp$ = perpendicular and $\parallel$ = parallel). Table 12.2 demonstrates the usefulness of such an orientation analysis. Figure 8.2a shows a highly oriented net of coronary capillaries surrounding cardiac muscle fibers. Figures 8.2b and 8.2c are two sections through this same net of capillaries. In the cross section (fig. 8.2b), we count 11 capillary sections, while we count $4\frac{1}{2}$ capillary profiles in the

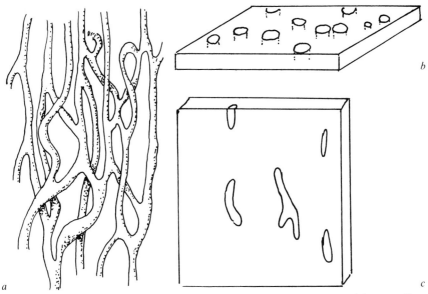

Fig. 8.2. (a) Stereogram of a coronary capillary net (from the subendocardial myocardium of a canine left ventricle). (b) A cross section through this net. (c) A longitudinal section through the same net.

longitudinal section (fig. 8.2c). Substituting these values into formula 8.3, we obtain

$$\Omega = \frac{P_{A\perp} - P_{A\|}}{P_{A\perp} + P_{A\|}} = \frac{11 - 4}{15} = 46.7\% \tag{8.4}$$

The numbers in this equation derive from two small pictures only. Therefore, the equation does not make a statement about coronary capillaries. Formula 8.3 can also be applied to skeletal muscle myofibrils and sarcoplasmic reticulum (the cisterns of which are known to be arranged in a basket-like fashion, but with the predominant direction parallel to the myofibrils). The following components of myofibers were estimated from figures 8.3 and 8.4: in longitudinal section (fig. 8.3) the profile count is myofibrils 92/1000 μm^2 and sarcoplasmic reticulum 230/1000 μm^2. In cross section (fig. 8.4) the profile count is myofibrils 570/1000 μm^2 and sarcoplasmic reticulum 346/1000 μm^2.

Formula 8.3 gives orientation values of 72.2% for myofibrils and 20.1% for sarcoplasmic reticulum from these data. To determine orientation, it is

A Guide to Practical Stereology 118

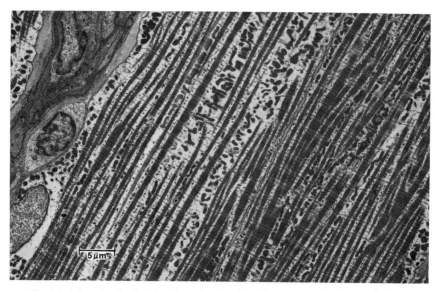

Fig. 8.3. A Longitudinal section through a skeletal muscle. The myofibrils and the cisterns of the sarcoplasmic reticulum are emphasized. (Micrograph provided by the courtesy of Dr. W.S. Tyler.)

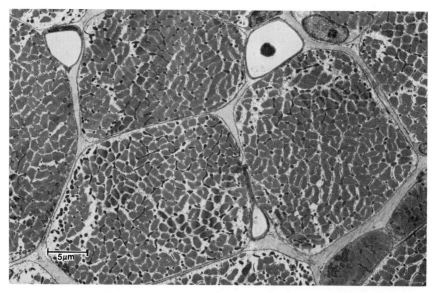

Fig. 8.4. Cross section through a skeletal muscle, again emphasizing myofibrils and cisterns of the sarcoplasmic reticulum. (Micrograph provided by the courtesy of Dr. W.S. Tyler.)

of utmost importance that these slabs are cut as perfectly parallel *and* as perfectly perpendicular as possible to the primary axis of orientation, an aim not always easily achieved. It becomes obvious then that for a completely parallel arrangement $\Omega = 1$, while it approaches zero in fully isotropic tissues.

Gradients of density (*nonhomogeneity*) exist in organs divisible into cortex and medulla. Density gradients may be combined with preferred orientation. A classic example is the cerebral cortex where the density and characteristic size of neurons are different in every layer and pyramidal cells are oriented perpendicularly to the surface of the brain. In the adrenal gland, the medulla is of isotropic construction. Not so the cortex: neighboring cell groups, cords and blood vessels of the zona fasciculata run parallel to each other and perpendicular to the capsule. In the kidney, the medulla consists of "straight" tubules and vasa recta, each of which runs parallel to its neighbors. Yet, very low power views show that on a large scale the "straight" tubules and the vasa "recta" are gently curved. In the kidney, a characteristic gradient exists in density and size of glomeruli, which are (a) absent from the subcapsular cortex down to a 1-mm depth, (b) numerous in the main part of the cortex, and (c) absent from the medulla. Further, juxtamedullary glomeruli are larger than subcapsular ones.

There are various sampling methods to deal with anisotropy and nonhomogeneity. When examining organs of small animals, it is recommended that one quantitatively analyze *entire* series of "sections" through the organ as well as study complete series of "sections" perpendicular to each other. For example, one should take a series of sagittal "sections" through the left and a series of transverse "sections" through the right kidney. Or when studying the cerebral cortex, evaluate a *few* "sections" *perpendicular* to the surface *plus* a *complete* series of "sections" *parallel* to the surface for each cortical area.

If the organ is large, this method is not practicable (see *Elias and Hennig* 1967). A human kidney, for example, is too large for serial sectioning. The following is a method recommended for quantitative kidney histology, which may be applicable to other organs. The entire kidney, after hardening, is divided into 1000–1500 little cubes, which are transferred into a liquid-filled dish. Then 50 cubes are extracted, by lottery, from the mixture and embedded; one slice through each is used for quantitative stereological counts and measurements. Each of the 50 slices is analyzed throughout using a sampling stage, and the results are treated collectively. By this method and by determining the volume of the organ as a whole, the size distribution of glomeruli, their number per unit volume of kidney, their total number in the entire kidney, the area of glomerular filtration surface for the entire kidney, the average length of the nephron, and so forth, can be determined.

Through dicing and mixing combined with a lottery, anisotropy has been artificially abolished.

The surface area of the cerebral cortex has been measured stereologically in many mammals (*Elias and Schwartz* 1971). Brains vary in shape from the very broad and relatively short brain of whales to the long and narrow brain of foxes. The shape of the neurocranium determines a preferred orientation of gyri and sulci. In every such case the brain must be divided into two hemispheres; one must be cut transversely, the other horizontally. Figure 8.1 (from *Elias and Schwartz* 1971) uses the example of a frontal and a horizontal slice through the cerebral hemispheres of a coyote on which arrays of equidistant parallel test lines have been superimposed in three directions of space, perpendicular to each other. Calculating the absolute value of $S = 2Pht$, one obtains the highest values of S for transverse test lines (A), a slightly lower value for vertical test lines (B), and the lowest value for longitudinal test lines (C). The arithmetic mean of the three values times 2, equals the absolute surface area of one hemisphere. (This method was developed in the late 1960s by *Elias* and co-workers before the brain was known to be bilateral.)

While all of our formulas are based on the theoretical assumption of accuracy, in any actual experiment irregularities of structure, accidents in placement of the test grid over an image, and deviations from the intended cutting direction introduce statistical errors that must be considered for each individual experiment (see Chapter 7).

Infinite Versus Finite Objects

An object of infinite extension is, for stereological purposes, one that reaches beyond the field of vision on all sides, as do slices from most organs when viewed through the microscope. When stereological counts are undertaken on an infinite object, they must be referred to a limited or finite test system. This problem has been presented masterfully by *Sitte* (1967).

Finite objects, such as a brain, viewed with the naked eye (figs. 3.9 and 8.1) can be tested by systems of points and lines that are of infinite or indeterminate extension. In this case, however, a complete series of slices of known thickness or slabs is needed, and the test points and test lines must be equidistant.

IX. Section Thickness and the "Holmes Effect"

Introduction

When dealing with true sections, i.e., plane cuts of thickness zero, such as the plane of polish in metallurgical specimens, or the opaque cut surface of a brain slice, stereology functions at its best. Such opaque materials, illuminated from above, presenting planar surfaces are easily treated by stereology. But histologists as well as geologists are confronted with translucent slices of finite thickness; so are metallurgists since the advent of transmission electron microscopy, because some metals are transparent to electron beams. Such preparations may present difficulties. Slices of measurable thickness are observed in transmitted light, in a direction perpendicular to their cut surface, yielding erroneous results.

This difficulty is called the Holmes effect after *Arthur Holmes* (1921) who, in his classic treatise on *Petrographic Methods and Calculations*, described it in the following words (p. 317):

> "The opacity of ore-minerals makes it impossible to detect the overlapping of boundaries within the thickness of a section, or to distinguish a small area on one surface from a larger area on the other. It is, therefore, always a maximum or cumulative area that is seen . . . , and consequently the measured intercept is always likely to be longer than it would be on the surface of a polished slab" (see fig. 9.1).

For example, when dealing with an opaque sphere, it is possible that a slice of the sphere is included in the translucent "section" and the larger perimeter becomes visible, while the smaller one remains obscured* (fig. 9.1).

Fig. 9.1. Portion of opaque spheres in a slice when viewed from above exhibit their largest diameter. Two small spheres, left of middle of picture, would partially overlap and appear to be fused.

---

* (The reader is reminded of slices of tomatoes and apples. The determination of size distribution of spheres is, therefore, sometimes called "the tomato salad problem.")

A simple way to avoid such difficulties has been used by one of us (Elias, unpublished observations) when estimating the size distribution of liver cell nuclei. These nuclei are spherical and exhibit a very large range of sizes having volumes from 1 to $6v$, the smallest being diploid, the largest dodecaploid. The easiest method to determine their size distribution appeared to be the production of very thick slices 25–35 μm thick, to stain the nuclei only and measure the diameters directly, considering only those nuclei whose equator was located within the slice (fig. 9.2a and b).

With the condenser high and the diaphragm open, the apparent diameter should come to a maximum within the slice, i.e., when focusing up, it should appear to become smaller. The same should happen when focusing down (fig. 9.2a and b). Nuclei whose diameter kept growing to either limit of the slice were considered to have their centers above or below the cutting limit and were not measured (fig. 9.2c and d). Such particles are sometimes called "truncated."

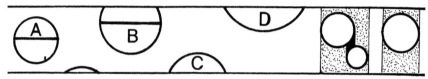

Fig. 9.2. The Holmes effect. See text.

A very similar technique, also not involving stereology, was used recently by *Ohno* (1980a) when measuring diameters of peroxisomes. When properly treated, these spherical organelles were opaque. Their apparent diameter is the larger of the two cut surfaces (fig. 9.1). *Ohno* (1978, 1980a) recognizing this difficulty, produced "thick" sections, up to 1 μm thick and observed them by high voltage TEM. He could determine the largest peroxisomal diameters; but, since it is not known, due to the opacity of the particles, whether the smaller appearing profiles derive from spheres located entirely within or at the boundary of the slice, only the upper size limit can be found by this method.

In a very lucid statement of the problem, *Haug* (1967) presented a number of formulas quoted from the literature, giving corrections for section thickness. Interested readers are referred to Haug's original paper. In general, it is desirable to avoid the effect of section thickness simply by producing very thin sections. This is often not practicable because the outlines of certain particles are not sharp enough for recognition in thin sections. But for sharply outlined particles, *Elias* has found (unpublished observations) that if the "section" thickness is one-sixth or less of the smallest diameter of the average particle, direct stereology is possible.

The Holmes effect not only interferes with size estimation, but also with the determination of volume ratio since particles overlap (fig. 9.2, right; fig. 9.3). If "sections" are thicker than $d/6$, the result of a point-hit count or of intercept measurements (linear scanning) will invariably show an exaggeratedly high volume contribution of the dispersed phase, until, at the section thickness of $\frac{5}{3}d$ it will appear as if the dispersed phase occupied the entire volume, if the particles are closely spaced (fig. 9.3) such as in the case

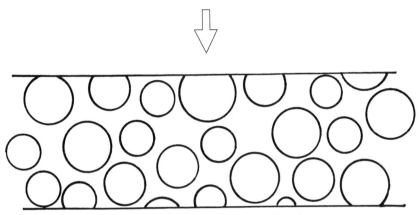

Fig. 9.3. When particles are packed closely in a thick slice, they may appear to occupy 100% of the volume. The matrix remains invisible.

of nuclei in many columnar and all pseudostratified epithelia.

Volume ratios are important in many biologic specimens. Examples are: nucleus/cytoplasm; glomeruli/cortical labyrinth; neuronal perikarya/neuropil; pulmonary alveoli/lung parenchyma; and blood corpuscles/whole blood.

The requirement of cutting sufficiently thin "sections" for light microscopy, previously possible only with extreme difficulty, can now be fulfilled with the advent of plastic embedding media and cutting with glass-knives. For many problems of new research in light microscopy, slices can be cut as thin as necessary. However, for many problems in pathology, this is not possible, since the investigator depends on old slide collections for the accumulation of a sufficient number of cases. The stereologist who investigates such slides must take section thickness and the Holmes effect into account. This is never easy when using high power light microscopy; and a general solution to the problem has not yet been found. Also, in low power (up to 5000×) transmission electron microscopy, conventional stereology can be practiced disregarding section thickness; but in high magnification electron microscopy the Holmes' effect remains a serious handicap since there are particles, such as ribosomes, synaptic vesicles, glycogen granules, microtubules, and so on, having diameters smaller than the thickness of "ultrathin" sections (50 nm). Synaptic vesicles are the most manageable of the above mentioned particles, since they are transparent to electron beams and have sharp outlines. But ribosomes and glycogen particles are electron opaque after the customary staining with heavy metals.

Several authors have treated section thickness and the Holmes effect mathematically and have devoted much space to it. For example, *Underwood* (1970) in his Chapter 6, *Projected Images* (pp. 148–194), has given it 46 pages. The most useful result, in our opinion, is his formula 6.5 on page 155 by which one can determine length in volume. The formula, slightly modified for easy understanding but in no way changed, is

$$L_V = \frac{1.7329 \, L'}{t \cdot A} \, mm^2. \tag{9.1}$$

L' is the combined apparent length of linear structures randomly distributed in space (see also Chapter 4 and fig. 4.1). Due to the enormous depth of focus of the transmission electron microscope linear structures obliquely oriented will be imaged sharply on the projection plane, even if they extend from the upper to the lower surface of the "ultrathin" section. This depth of focus of the TEM is due to the extreme narrowness of the aperture, a feature necessary to enhance contrast.

When considering this specific problem of length/volume determination, one must be aware of the fact that we are dealing with extremely thin features (microtubules, protofibrils, microfilaments, etc.) and that the necessary magnification is so great that only a tiny fraction of a cell is shown in a single field of vision. This field is not necessarily representative of the entire cytoplasmic extent, for there are areas in the cell where these thin linear structures must be absent, as in the Golgi apparatus, in lysosomes and other portions of the cell.

Thorough sampling is necessary. This consists of the following considerations:

1. Use the largest possible negative size.
2. Take micrographs from many sections of the same kind of cell in order to achieve randomness of cutting directions.
3. Sample the *entire* cell, including those portions in which the linear

structures are entirely absent so as to be able to make a statement such as 22 inches of microtubules per cubic millimeter for the entire cell (now, these 22 inches are nothing but a humorous guess, but not impossible if one considers that the ductus epididymidis is estimated to be $\frac{1}{3}$ mile long, and the total length of the glomerular blood channels in one kidney of a 10-year-old boy was found to be 7 miles long, and that microtubules are much thinner and more densely packed).

As in the determination of quantitative data on renal glomeruli, the entire kidney, including the medulla as well as cortex, had to be sampled and treated collectively, and so it must be with components of a cell.

Weibel (1979) in his volume I devotes subchapter 4.3.4, 11 pages (pp. 139–150) and in Volume II Chapter 4 (pp. 105–139), 34 pages, to section thickness and the Holmes effect. Nevertheless, he does not achieve a satisfying result. He introduces shape coefficients which refer only to spheres, disks and rods, and he admits that this "first model can, in fact, only be expected to apply if the structural elements are very loosely arranged, i.e., if the V_V is very small; in this case the chance that profiles generated by a thick slice overlap, is also small . . . (and) . . . that the elements are assembled by a true random process, a condition often not fulfilled."

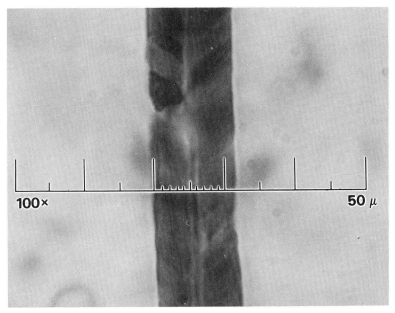

Fig. 9.4. A light micrograph of a Small-fold showing that the "section" thickness is exactly as the setting of the microtome specified.

The present authors have, themselves, struggled unsuccessfully with section thickness, especially when particles were closely packed and anisotropically oriented. The conclusion is that at this time, no generally valid mathematical solution for the Holmes effect has been found. It still remains necessary to search for a special solution to each specific problem. One important factor in any such attempt is the necessity to determine the thickness of the "section." Various methods exist to do this. We have discussed thickness determination in Chapter 5. Figure 9.4 shows, again, a so-called Small-fold measurement with an oil-immersion objective.

X. Automated Methods in Stereology

Automated Image Analysis

Advances in television scanning and electronic techniques have enabled image analysis to become fully automatic. In essence, automated image analysis is performed by a high-resolution television camera that scans either a histologic slide (with the aid of an optical microscope) or a photograph (with the aid of an accessory lens and illumination), and then converts electrical analog signals for each point on the screen into digital signals, and stores or processes the digital data.

Variations in software and hardware for digital data handling differentiate the five commercial and custom-designed image analyzers. For example, two commercially available analyzers, the Quantimet 720 (*Fisher* 1971) and the Omnicon (*Morton and McCarthy* 1975), differ considerably in their software and hardware applications to data processing. However, all of the systems operate on the same basic principle of manipulation of digital data from analog signals generated by detected points on a television screen.

Naturally, preparation of the specimen to enhance contrast in the structure to be detected is a prerequisite to successful differentiation by any system. For example, we have used a variety of silver stains to stain interalveolar septa of lung tissue a dark black color, while others (*Langston and Thurlbeck* 1978) have overstained the same type of tissue with eosin and used a green optical filter. Histochemical stains have also been used, for example, to differentiate arterial blood vessels, using the ATPase reaction (*Herrmann et al.* 1979) and type 2 pulmonary cells, using the lactate dehydrogenase reaction (*Sherwin et al.* 1973). The greatest potential for advances in biologic application of automated image analysis lies in the area of selective stains for cells and cellular contents, since the machine is primarily limited by the quality of specimen preparation.

Assuming that the preparation of the biologic specimen is satisfactory, there still are potential problems with commercially available image analyzers. In the field mode, where all features in the field are treated at the same time, spurious intercepts have resulted in an underestimation of interalveolar wall distances as compared to manual methods (*Hyde et al.* 1977; *Langston and Thurlbeck* 1978). Both investigations attributed the major source of spurious intercepts to pulmonary capillaries and cells within blood vessels. Spurious intercepts can be generated by inclusions of dirt, tears in tissue, scratches on the glass surface of histologic slides, or other artifacts (fig. 10.1). It is crucial that these artifacts be meticulously avoided. However, estimates of volumetric density in the field mode appear to be very close to results using manual methods (*Hyde et al.* 1977; *Stinson et al.* 1977). Both of these studies required meticulous calibration of the threshold setting for the detector. If there is variability in the threshold setting from day to day, the respective detected volume densities will show an increased variability.

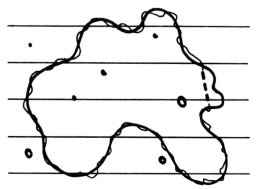

Fig. 10.1. Spurious intercepts are generated using automated image analysis by inclusions in the main structure of interest. Semiautomated methods have the inherent problems of reproducibly tracing complex structures and returning to the exact starting trace point. The machine automatically completes the feature from the end to starting trace point with a straight line. [Reproduced with permission from Elias, H. and Hyde, D.M.: Elementary stereology. Am. J. Anat. *159:* 411–446 (1980).]

In the feature mode, where individual features within a field are measured, there is much greater flexibility. In most systems the feature mode requires more computing capability; hence, the machine cost is greater than for a machine with the field mode alone.

There is a dearth of investigations comparing measurements of individual features by image analyzers and manual methods. However, the results published to date appear promising. *Stinson and Sporn* (1977) compared counts of urinary bladder epithelial nuclei by an image analyzer and manual methods and found no significant difference between methods, provided nuclear images were not touching. Many of the accessory features of image analyzers, such as a light pen (for operator interaction with the displayed image) and a two-dimensional amender (which allows an operator to dilate or erode image output from a detector) make automatic separation of features possible. A recent study of growing rat lung involved detailed feature analysis using a Quantimet 720 image analyzer interfaced to a PDP 11/10 computer (*Keller and Burri* 1981). In this study the two-dimensional amender was used to simplify the image for a count of air spaces. Alveolar counts attained using this method correlated well with previous investigations of the growing rat lung.

With the assistance of an on-line computer, the capabilities of image analyzers are almost limitless, provided the specimen is optimally prepared for the detector. However, there are potential disadvantages when automated image analyzers are not coupled to a computer. Such a large volume of data is generated that the time advantage gained in automated counting is lost to an enormous expansion of data. Thus, in our opinion any system that ultimately benefits the investigator must have data reduction, statistical summary, and extensive memory storage capability.

We will now provide a description of our approach to determining the surface density and volume density of alveolar tissue and airspace using a Quantimet 720 image analyzer. As stated previously, a silver stain is used to darken interalveolar septa. For the light microscope connected to the image analyzer a $4\times$ objective and a $5\times$ ocular are used, which gives a screen picture point width of 2.4 μm. Before the slide is placed on the microscope, screen detection is adjusted so that it is even across the field. Next, a standard specimen (we use a grid pattern) is used to adjust the detection level to a predetermined value that gives optimum detection of interalveolar septa. This procedure does not compensate for inadequately stained slides, hence the operator must exclude any poorly stained slides to avoid generating false data. This requires a subjective decision by the operator for each slide.

There are two basic programs: a chord-sizing program that is used to determine the mean linear intercept of airspaces, and a pattern-recognition program that counts the number of alveoli and determines the volumetric densities of alveoli and alveolar ducts. The chord-sizing program uses 14 size classes to accumulate chord lengths of airspaces. These data along with slide identification and a variety of other features of interest concerning the experimental animal are recorded on magnetic tape and subsequently processed in a computer. The computer programs used to process the data are (a) a data read program that creates a file from tape sequence, (b) a data editor program that checks the new file record by record for correct format and allows for data editing, (c) an image analyzer program that provides basic stereological data of volumetric densities of airspace and interalveolar tissue and the surface density of alveoli and (d) a chord distribution analysis that determines the mean linear intercept from a cumulative frequency distribution of chord lengths.

Since program 3 uses conventional stereological formulas for calculation, only the formulas used in program 4 will be given here. A distribution of percent cumulative frequency of airspace chords in 14 size groups is computed for each lung. A two parameter cumulative probability law of the form $F(x) = P(X \leq x) = (1 - e^{-\alpha x})^\beta$ was found to provide the best fit to that distribution. The parameters α and β were estimated using a combination of minimum modified chi-square (*Cramer* 1946) and the Newton-Raphson procedures (*Fröberg* 1965). The estimated mean chord length in x was obtained by calculating the area above the curve using the sixth order Newton-Cote formula for numerical integration (*Fröberg* 1965):

$$\xi(x) = \int_0^\infty [1 - F(x)]\, dx \tag{10.1}$$

The variance of x was determined by the following formula:

$$\hat{\sigma}_x^2 = 2 \int_0^\infty x[1 - F(x)]\, dx - \left[\int_0^\infty rF(x)\, dx\right]^2 \tag{10.2}$$

The chord distribution method has been found to be very sensitive to changes in the architectural arrangement of alveolar airspaces.

The other basic program that utilizes pattern recognition necessitates the use of a light pen. Each field must be edited using a light pen to close alveolar spaces off from alveolar duct and sac lumina. Subsequently alveoli can be treated separately from alveolar ducts and sacs by areal sizing. Hence, it is possible to obtain alveolar duct and alveolar volumetric densities using this program. Because the programs are over 100 pages in length, they are not included in this book.

Semiautomated Image Analysis

A potential compromise between the human ability to recognize and classify and the computer's ability to sort according to dimensional measurement and to reduce data is the graphic digital tablet connected to a computer. There are a number of commercially available graphic digitizers; all of them require the use of a programmable calculator or access to a computer. Traced coordinates are classified and digitized for subsequent processing by a computer. In most custom-designed systems, the small computer is the most expensive part of the system (*Dunn et al.* 1975), yet the systems as a whole are relatively inexpensive.

A small on-line computer can serve multiple functions: as an interactive source of manual data recording via a peripheral keyboard; as a data processing device; and as a statistical data summarizer (*Raetz et al.* 1974). One of the authors (DMH) has found the on-line computer a great tool because all stereological formulas can be stored, then recalled as appropriate to treat selected data. Coupled to graphics terminals, high-speed printers, and a variety of other peripheral devices, the system gains a wide range of application and flexibility.

Commercially available graphics tablets use voltage, sound waves, magnetic pulses, or mechanical means to record coordinates in X and Y directions. To date, no study has been published, which compares the tablets available commercially; most manufacturers claim a precision of 0.1 mm for coordinate specificity.

One potential problem with graphics tablets for routine stereological measurements is that inherent in the reproducibility of tracing complex structures and returning to an exact starting trace point (fig. 10.1). Most systems automatically complete a feature from the end to starting trace point with a straight line.

Only a cursory comparison has been completed between results obtained using a graphics tablet coupled to an on-line computer and those obtained when manual counts are entered on a computer; however, volumetric densities were comparable in variance, whereas surface density was more variable in the tablet/computer system. In addition, the time per field was slightly less using the manual-count method than using the tablet/computer system. *Mathieu et al.* (1981) have recently confirmed these observations in a comparison of manual, automated and semiautomated image analysis methods using biologic specimens. Based on this evidence, computer analysis of manual counts is recommended for routine volume, surface, and numerical density determinations. For more complex measurements such as diameters, axial ratios, and X-Y coordinate identification of features, however, the digital tablet/computer system is superior. Certainly, if an investigator is planning to make a series of simple and complex measurements of a field, a digitizer tablet/computer system is the method of choice.

A very important application of the tablet/computer system is its use in three-dimensional reconstruction. Because of the relationship of mean caliper diameter to numerical density, three-dimensional reconstruction is a very powerful biologic tool for estimating cellular numerical density (see Chapter 11). There are numerous examples in the biologic literature of applications of the digitizer tablet/computer system to three-dimensional reconstruction. Many of these applications have been directed to reconstruction of neurons and neuron assemblies; *Macagno et al.* (1979) have recently reviewed these applications.

One interesting application that should be developed using the digital tablet/computer system is the calculation of axial ratios from traced coordinates. As evidenced by discussions in Chapters 4 and 5, the axial ratio measurement is tremendously useful. However, one must usually use a concentric circular grid to determine the ratio (fig. 10.2). This is a tedious measurement and with fatigue can lead to potential error.

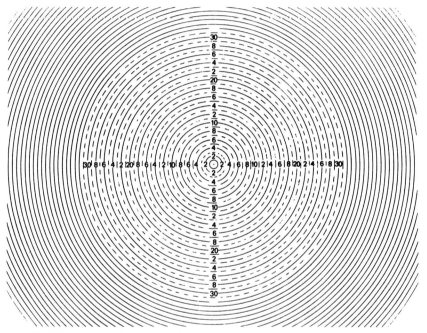

Fig. 10.2. A concentric circular grid used to measure axial ratios of profiles. Odd numbered circles are dashed lines.

Assume that a coordinate file for each profile traced has been created. Instruct the computer to find the maximum profile diameter (D_{max}). Then instruct it to find the normal (perpendicular) maximum diameter to D_{max} and the axial ratio is obtained. To accomplish the above stated objective, we must first look at our coordinate file. It is a series of x and y coordinates, usually about 200 to 300 in number, describing the particle profile. By transformation of coordinates (rotation of the axes), the minimum and maximum x values can be evaluated for each rotation. The formulas for rotation of coordinate axes are:

$$x' = x \cos \theta + y \sin \theta \qquad (10.3)$$

$$y' = y \cos \theta - x \sin \theta \qquad (10.4)$$

where θ is the increment of rotation in radians and x' and y' are the new coordinate locations relative to the rotated axis. Rotation through 180° in 10° increments about the y axis is accomplished while x is held constant. Two arrays, one for x_{min} and one for x_{max} of 18 × 1 dimensions can be created. For each rotation, each of the coordinates for X_{max} and X_{min} is evaluated and the greatest and least X values are placed in the appropriate arrays. When the rotations are completed, the arrays are evaluated for D_{max}, which is the largest profile tangent diameter. Then the values for X_{max} and X_{min} corresponding to the rotation perpendicular to D_{max} are found and the profile axial ratio calculated. To our knowledge such a program has not been written. Hopefully the previous discussion will be instrumental in its development.

XI. Three-Dimensional Reconstruction and Other Ancillary Stereological Techniques

Introduction

If a structure is solitary or if its shape is complicated, geometric-statistical methods are of little use. For determining the shape of such objects, reconstruction from serial sections is the best method available at this time.

Many researchers present incorrectly written results of their work, claiming that they have investigated serial sections. Unfortunately, no human being is capable of conceptualizing correct three-dimensional images of specific objects from visual observation of such sections. The reason for this difficulty, which seems to be universal among morphologists, is the inability to memorize successive images that look very similar and also the inability to retain their succession.

The identification of a three-dimensional structure from *single* sections using the "commonsense approach," has been discussed. This chapter deals with methods of comprehending complicated shapes and problems of continuity versus discontinuity (as for example in a problem involving the congenital discontinuity of a duct system, as in biliary atresia).

A method that can provide insight into such problems without reconstruction is the Leporello-register method. (The word derives from the portrait catalog of Don Juan's girlfriends arranged by his man-servant Leporello in accordion fashion). The Leporello method involves taking photographic negatives of an uninterrupted series of sections and mounting the prints on a long, folded strip of material so that it becomes possible for the investigator to view long series of sections simultaneously and side by side in proper succession.

Figure 11.1 shows three Leporello series of serial sections through the liver region of a young garter snake and of two shark embryos. These examples were chosen because they happen to be the shortest series in the senior author's collection. The Leporello method facilitates proper alignment of successive sections. It can precede actual reconstruction, as it has when several long series of sections were photographed in a single day in out-of-town collections. The precision of reconstruction depends on adjacent

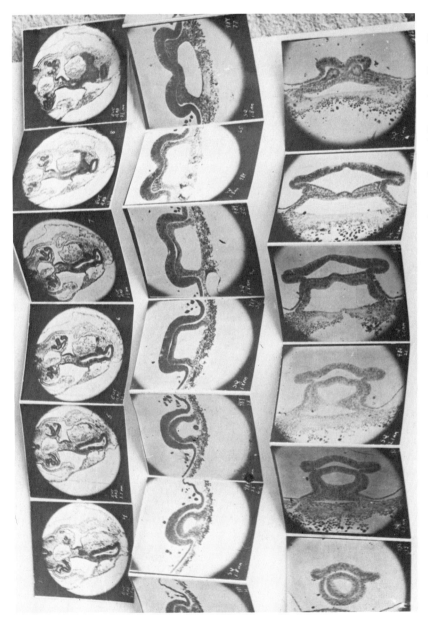

Fig. 11.1. Three Leporello-registers of early vertebrate embryos emphasizing the midgut region and liver primordium. *Upper row*: *Thamnophis sirtalis* (garden snake); *middle row*: *Squalus acanthias* (shark); and *lower row*: *Scillium canicula* (dog fish).

section orientation, knowledge of section thickness, and some method of visualization of the reconstructed image. The problem of adjacent section orientation is bypassed using the technique of photographing the block face after each section is cut (cinematography). The application of this technique may provide a great deal of knowledge concerning the subgross anatomy of organs.

Another technique, stereophotogrammetry, uses displacement of an object relative to a photographic source (parallax) to measure height in the third dimension. Although this technique is well established in aerial geography research, its application to histology is practically unknown. The word morphometry was first coined and used by geographers and brought over into the biomedical field by *Weibel* (1963).

One final technique that warrants mention in this context is the drawing of a conceptual image derived by observing many sections through an object (a stereogram).

Three-dimensional Reconstruction

D. Born (1883) is credited with the invention of the reconstruction method. Correct orientation and "registering" of successive serial sections presented difficulties. These were overcome in 1886 by *Kastschenko* and refined by him in 1888. He trimmed a paraffin block close to the object and painted it with a black lacquer, then surrounded the painted block externally with more paraffin. Only then was the object cut. The lacquer produced black, rectangular frames around the sections, which permitted excellent alignment. *Peter* (1899) refined the method and in 1906 wrote a book about the reconstruction methods then known. In 1922, *Peter* devised methods for reconstruction in planes oblique to the cutting direction. These planes of orientation (Definierebenen) can be used only at very low magnifications and if reconstruction has been planned before cutting.

A few methods that permit good alignment at high magnification and on serial sections without orientation planes are described below. Orientation (registering) of adjacent sections is absolutely crucial because all other facets of the reconstruction and three-dimensional morphometry depend on it. Since *Gaunt and Gaunt* (1978) have reviewed methods of section orientation in Chapters 2 and 3 of their book *Three Dimensional Reconstruction in Biology*, only a few methods currently used in our laboratories are presented.

Three levels of reference marks can be used to assist in adjacent section orientation. At level 1 the block face is trimmed on one corner to insure that the section is placed with the correct side up each time. This is for subgross orientation. Black silk suture threads (8-0 to 10-0) are used for microscopic

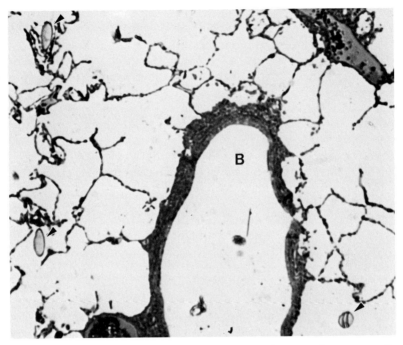

Fig. 11.2. Three black nylon sutures (9-0) (arrows) are sewn into parenchyma surrounding a nonalveolar bronchiole (B) of a cat lung prior to embedding in araldite. Subsequently, the sutures are used for adjacent section orientation (1-μm section).

alignment (level 2) using a low-power objective lens. Usually three threads are sewn adjacent and perpendicular to subgross features of interest. After they are embedded and sectioned in plastic, these black threads are visible in sections adjacent to the feature of interest (fig. 11.2). For most structures 50 to 500 μm in diameter, orientation at levels 1 and 2 can be very adequate; however, for the reconstruction of cell nuclei, lines can be drawn between nuclear profiles for very subtle adjacent-section best fit.

Although this technique has not been applied to thin sections in the transmission electron microscope, we can envision a closer placement of black silk threads (about 1 mm apart) for low magnification orientation. It becomes obvious to electron microscopists that this would necessitate cutting a block face of about $1\frac{1}{2}$ mm^2, mounting sections on slot grids, and correcting precisely for the degree of orientation change with increasing magnification. Intracellular organelles could also be used for orientation at level 3 by adjacent-section best fit.

Knowledge of section thickness is often very important. It is not important if continuity is the only aspect to be investigated. For most commercial ultramicrotomes, the instrument thickness setting corresponds very closely to the actual thickness (*Loud* 1978; *Ohno* 1980b). However, it is always wise to check section thickness by the methods described in Chapter 9.

Sketching serial sections of a structure onto wax or glass plates is a common approach to three-dimensional reconstruction. The magnification of the projected structure is determined by the ratio of the plate thickness to the actual section thickness. In the wax plate method, the images are projected on wax plates and cut out; successive cut-out tracings are pasted together, thus creating a model of the object. The glass plate method involves projecting successive slices on glass plates or plastic sheets and tracing them with ink. These transparent plates are then stacked, and the three-dimensional image is visually apparent. The senior author has used the following two methods of alignment for long series of sections through embryos (up to 200 slides with an average of 12 sections on each). Such sections, which are found in existing collections of embryos, are never properly aligned, although their alignment is not difficult at all, even at medium power (up to $20 \times$).

Table Projection. An inclinable, monocular microscope is placed in a horizontal position on a table, conveniently raised on a stand (fig. 11.3). A low voltage microscope illuminator is placed on the level with the barrel of the microscope. An inclinable mirror is mounted in front of the eyepiece, and the microscopic image is projected on the table. By the proper choice of objective, eyepiece, height of stand, and distance of mirror from the exit pupil of the eyepiece, an image can be created that measures about 6×8 cm ($3\frac{1}{4} \times 4$ inch), the size of commercially available lantern-slide cover glasses. While figure 11.3 was painted, a $15 \times$ eyepiece was used to create a large image. When traced on lantern-slide cover glasses, the image must be smaller. A 3-5 $\times$ eyepiece is then preferable.

A section is selected from near the middle of the organ of interest. The image is projected on a sheet of white paper or cardboard (whichever is preferred) on which the outline of the glass plate has been traced.

Then, on a piece of white (nonreflecting) cardboard, essentially larger than the glass plates, build a frame with three sides of thick cardboard into which you can insert a standard size glass plate. That part of the cardboard base outside the frame is weighted to keep it steady. Now, the image is centered on the plate area. The first glass plate is placed there and the outlines of the interesting structure on that plate are traced, possibly with bright, opaque ink. Every kind of structural element, such as the bile duct and gallbladder, the stomach epithelium, the duodenal epithelium, the outline of the liver, blood vessels, and so on, can be traced in a different color.

Three-dimensional Reconstruction and Other Ancillary Stereological Techniques

Fig. 11.3. Table microprojector used for serial section reconstruction. Oil painting *My Three Assistants* by Hans Elias, 1956.

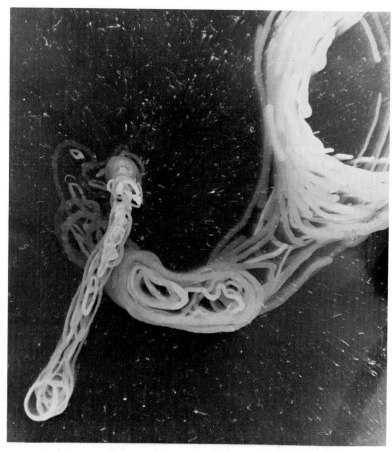

Fig. 11.4. Glass plate model from a 16-mm human embryo, emphasizing the stomach (*right*) pylorus, proximal duodenum (*left*), and the extrahepatic biliary system. [Reproduced with permission from Elias, H. and Hyde, D.M.: Elementary stereology. Am. J. Anat. *159*:411–446 (1980).]

After the first tracing is completed, the same place in the next section is searched for in the microscope and centered. The previous glass tracing remains in the frame and the cardboard frame moved until it aligns with the new image. Insert a new glass plate and proceed in the same manner as for the previous plate until the end of the organ is reached. Then the process is repeated, beginning on the other side of the section traced first. Every glass plate must be numbered. After the series of glass tracings is complete, they can be stacked and, the model yields a good image of reality (fig. 11.4).

The Expanded Leporello Method. Place the first glass plate on the photograph with which you wish to begin. When proceeding to the next picture, lay a blank glass plate on that picture and the previous tracing on top of it. Move it around until it fits. Remove the old tracing and trace the new image.

These methods of reconstruction have certain advantages and disadvantages that must be weighed in selecting a technique. Wax-plate models can be viewed from every angle, which is not possible with glass-plate models. However, wax-plate models present serious difficulties in reconstructing branched or highly lobed structures. For example, a branched duct system cut by a plane (fig. 11.5) represented by the heavy straight line, yields several discrete profiles (fig. 11.5). When cutting out wax plates, bridges must be left (dotted stripes in fig. 11.5) to keep individual profiles in their proper relationships. With the glass-plate method, these profiles are fixed in their positions by the rigidity of the glass.

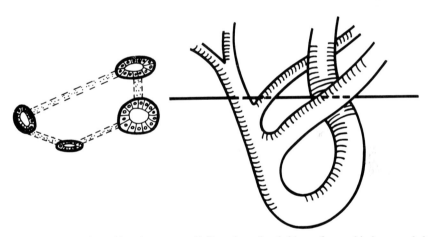

Fig. 11.5. A branching duct system (*left*) to show that bridges of some kind are needed to keep profiles of parts in proper relationship (*right*). [Reproduced with permission from Elias, H. and Hyde, D.M.: Elementary stereology. Am. J. Anat. *159:*411–446 (1980).]

The interpretation of glass-plate models requires a certain degree of spatial imagination. An impression of depth can be created if bright, opaque colors are used, and if the model is viewed against a black background and illuminated obliquely from above. A little dust on the glass plates enhances the effect (fig. 11.4).

The two methods also differ in the amount of time required to complete a reconstruction. A wax-plate model may require as much as 3 weeks to produce, since it involves projection, tracing, cutting out, welding the cutout plates together, removing the bridges, and smoothing out the edges, whereas a glass-plate model of the same object can be made in about 3 hours.

*Computer-aided Reconstruction**

In contrast to these traditional methods of reconstruction, computer aided reconstruction provides some distinct improvements. These include the time required to record the serial profiles of a structure, the ability to view the reconstructed image from all perspectives, and the capability of making numerous rapid, yet precise three-dimensional measurements on the reconstructed image. Since an unbiased estimate of the three-dimensional measurements of a population of bodies is desired, area-weighted random sampling is used to select the reconstructed bodies from their midsection (see Chapter VII). Inherent in this sampling method is the assumption of independence of the size of the body from its location. The previously described methods of serial section orientation are used to bring oriented profile tracings of known thickness to the computer. A digitizer tablet is serially connected to an on-line computer (DEC LSI 11/23) to create a Cartesian (X, Y) coordinate data file of each serial profile at a resolution of 0.1 mm between coordinates. The digitizer is used in the continuous sample mode and all coordinates are referenced to a fixed origin. By entering all serial tracings in their aligned format with each section coded for a new Z coordinate, a three-dimensional (X, Y, Z) file is created for each series. The subroutine called GETSET in Appendix 1 allows up to 500 coordinates per section, but there is no restriction on the number of sections.

After all sections are digitized, they can be viewed from any perspective with a program that uses the coordinate files to reconstruct the three-dimensional image. Figure 11.6a shows the reconstruction of a pulmonary alveolus as it was sectioned from apex to base with no rotation, while figure 11.6b shows the same alveolus rotated 75° on the X axis. As with any serial profile reconstruction, hidden lines assist in the three-dimensional impression of the image. Motion or real-time rotation of an image in this application also enhances the three-dimensionality of an image. However, this capability is only provided on large and relatively expensive computers. Fortunately, all of the programs listed in Appendices 1 and 2 can be run on a microprocessor with a memory of 48K Bytes.

---

* We thank E. Reus, N. Tyler and W. Tyler for writing the computer programs and assisting in writing this part of the chapter.

Three-dimensional Reconstruction and Other Ancillary Stereological Techniques   143

Fig. 11.6. A distal alveolus, reconstructed from 20 μm serial sections. (*a*) viewed as it was sectioned, apex to base and (*b*) viewed after a 75° rotation about the *X* axis.

Woody et al. (1980) determined morphometric parameters from reconstructed serial sections. They described the estimation of mean caliper diameter ($\bar{D}$) by random rotation of coordinate axes. Unfortunately, they did not report their algorithm for the rotation of coordinate axes. However, they did include a comparison of the errors introduced by the Holmes effect and lost polar caps using a spherical model. Both of the errors are reduced by increasing the number of sections and the number of points per profile.

In the following presentation we describe an approach developed independently of Woody et al. (1980), to estimate the mean caliper diameter ($\bar{D}$) of a reconstructed structure using a computer. To conserve memory, maximum and minimum X values at each rotation are computed for the first section and these values are stored in two arrays. The next section updates the minimum array since Z becomes more negative with each additional section. When all sections have been evaluated the final arrays are used to compute $\bar{D}$.

To rotate coordinate axes, draw three orthogonal axes X, Y and Z (fig. 11.7). For rotation around the X axis, both Y and Z are rotated to Y' and Z' (fig. 11.7a) using the following formulas:

$$Y' = y \cos \theta_1 + Z \sin \theta_1 \qquad (11.1)$$

$$Z' = Z \cos \theta_1 - Y \sin \theta_1 \qquad (11.2)$$

For rotation around the Y axis, both X' and Z' are rotated to X'' and Z'' (fig. 11.7b) using the following formulas:

$$X'' = X' \cos \theta_2 - Z' \sin \theta_2 \qquad (11.3)$$

$$Z'' = Z' \cos \theta_2 + X' \sin \theta_2 \qquad (11.4)$$

For each section, both X and Y are rotated through 180° by incrementing X 10° and rotating Y in 10° increments through 180°, then incrementing X an additional 10° and rotating Y through 180°. At each individual rotation all coordinates are evaluated to determine X_{max} and X_{min}. This process of rotation is continued until 18×18 rotations have been completed for each section and there are two arrays, one for X_{max} and the other for X_{min}, both 18×18 in size.

From these two arrays, 324 individual caliper diameters (D) are calculated. Since the density of rotation is not constant using polar coordinates, the mean caliper diameter ($\bar{D}$) is weighted according to its Y axis rotation by the following formula:

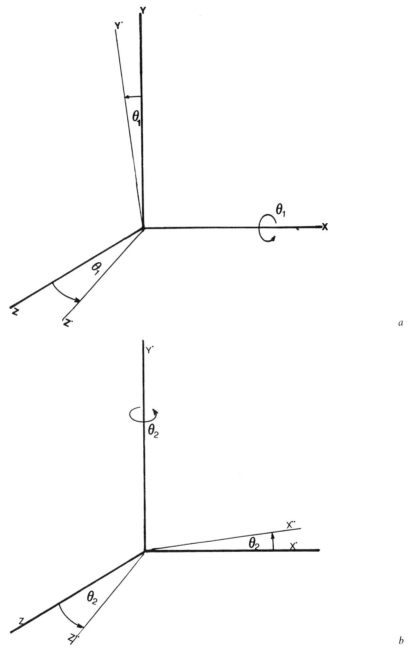

Fig. 11.7. Three orthogonal axes X, Y and Z are used to estimate mean caliper diameter. The program rotates around the X axis (a) and around Y axis (b).

$$\bar{D} = \frac{\sum_{j=1}^{m} \sum_{i=1}^{n} D_{(\theta i, \phi j)} \cdot \mathrm{Sin}\,(\phi_j)}{\sum_{j=1}^{m} \sum_{i=1}^{n} \cdot \mathrm{Sin}\,(\phi_j)}, \tag{11.5}$$

where n is the number of X rotations, m is the number of Y rotations, $D_{(\theta i, \phi j)}$ is the caliper diameter at any X/Y rotation, and ϕ_j is the y axis rotation.

With a spherical model, a ratio of section thickness to true D of 0.1 or 10 sections per body was found to give an estimate of D of less than 5% error. In practice, a ratio of 0.05 is used, which we believe minimizes errors due to the Holmes effect and lost polar caps. The use of D to estimate numerical density of biologic structures is presented in Chapter V.

Another method of reconstruction from serial sections is serial section cinematography, where a motion picture camera is focused on the surface of the paraffin block in which the object is embedded, mounted in a microtome. After it is appropriately stained, its surface is photographed on one frame of the film. This process is repeated every time a slice is cut from the block (*Hegre and Brashear* 1946; *Postlethwait* 1963). In this manner a photographic record is obtained of serial sections of the object under study. The photographic sections are in perfect register with each other and free from distortion. Reconstructions can then be made by projecting one motion picture frame after the other on wax or glass plates, or by digital entry into a computer.

Many other methods exist for serial section reconstruction. A complete treatment of these methods is beyond the scope of this chapter. Reconstruction by serial-section cinematography is useful in resolving the topologic question of continuity versus discreteness. A preliminary stereological approach to the problem of continuity was made by *Hennig and Elias* (1966).

Computer Aided Tomography (CAT) has had a significant impact on radiology and nuclear medicine. Using existing x-ray technology and more recently developed computer signal processing, it has allowed radiologists to view slices through the body in which organs are unencumbered by the shadows of other organs and tissues (fig. 11.8). In nuclear medicine images are formed of organs and tissues containing gamma-emitting radionucleides; however, this later method has a much lower resolution than x-rays because collimating the gamma source to improve resolution reduces detector efficiency.

Stereological applications to CAT images can provide valuable morphometric measures of organ volume and dimension. A recent report by *Heymsfield et al.* (1979) used CAT images to estimate liver, kidney and spleen volumes. In essence, their apparatus and method estimated the area A_i of each "section" through the organ. The sum of all areas multiplied by the

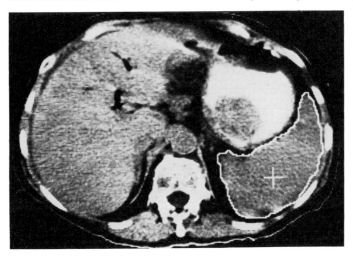

Fig. 11.8. A computer aided tomographic image of a human patient at the level of the first lumbar vertebra. The slice image, viewed from the caudal aspect, shows the spleen encircled by a white line and marked with a +. This line, produced with a "cursor" is larger than the real trace of the organ outline since the cursor converts a smooth line, drawn by an observer, into rectangular coordinates. The stomach, partially filled with contrast media, is ventral to the spleen and the liver occupies most of the opposite side of the abdomen (courtesy of Dr. Michael Federle, San Francisco General Hospital.)

"section thickness" (distance between images), equals the volume of the organ (formula 3.1). Since these images by nature are shadows of semi-translucent organs, the Holmes effect plays an insignificant role since the outline is traced through the middle of its penumbra. They found this method to be accurate to within 3% to 5% of actual organ volume using human cadavers. *Moss et al.* (1981) have also used CAT images of experimental dogs to determine the volumes of liver, spleen and kidney. They sacrificed the dogs, checked the organ volumes directly by water displacement, and reported an accuracy of within 5% for the CAT volume estimate. *Federle and Moss* (1981) recently used this technique to estimate the volume of spleens of human patients and found it of great diagnostic value (fig. 11.8). *Oldendorf* (1980) in his book, *The Quest for an Image of Brain*, presents an overview of CAT imaging techniques, while *Scudder* (1978) reviews the basic physics and mathematics underlying the production of reconstructed tomographic images.

Stereo-photogrammetry

Stereovision depends on parallax, the apparent difference in an object caused by viewing it from two different points (fig. 11.9). It is parallax that allows humans to visualize depth. This same principle can be applied to microscopic structures by tilting the specimen to generate parallax. *Boyde* (1973) pioneered applications of photogrammetry using scanning electron

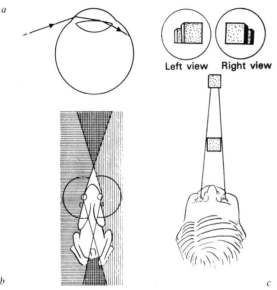

Fig. 11.9. Panoramic versus stereoscopic vision. [Reproduced with permission from Elias, H.; Pauly, J.E., and Burns, E.R.: Histology and human microanatomy; 4th ed. (Wiley, N.Y. 1978).]

microscopy. *Boyde and Ross* (1975) reviewed the limitations of scanning electron microscopic photogrammetry, reporting the appropriateness of photogrammetric measurements of solid tissues such as bone. We concur in their recommendation of this method to solid tissues, but for most organs and tissues three-dimensional reconstruction is more accurate and is preferred. Another excellent application of stereophotogrammetry is high voltage electron microscopy. In this application sections from 1 to 5 μm can be observed in stereovision and height measurements made of subcellular detail. Since *Gaunt and Gaunt* (1978) have devoted an entire chapter (8) to stereo-related methods of photogrammetry, readers are referred to their excellent discussion of this topic.

XII. Applications of Stereological Methods to Various Organs

Introduction

In this chapter we have selected three divergent organs as examples of how to perform a stereological study. As in any experimental investigation posing the proper questions is the key to the selection of the best technical method. Stereology is ideally suited to experimental evaluation of the toxicologic effect of a variety of agents on organs and tissues. It is equally well suited to the evaluation of other pathologic changes. However, even in such cases the economic considerations of a study dictate that the investigator examine structures that are of the greatest functional importance and that they be evaluated according to their priority of importance.

In any experimental study the organism is the primary sampling unit. Variation within a species can be estimated and sampling within the organ adjusted to reach an acceptable level of error (see Chapter 7). However, variation between individuals is influenced by basic biologic variation and a variety of other factors due to the differential biologic response to experimental or pathologic perturbation. Hence, it can prove of great usefulness to run a pilot study in a new investigation in order to estimate the feasibility of conducting any proposed investigation.

A thorough literature search can be extremely valuable in predicting experimental number. Yet, the investigator must guard against taking published opinions, often confused with "facts," too seriously. Let nature speak to you. Remember, printed statements are frequently biased. One should always be sensitive to the potential correlation of structure and function within individuals. In preliminary studies one of the authors (DMH) has found a stepwise regression procedure of structure on function to be very useful in explaining observed variation in pulmonary function measurements using stereological measurements of lung tissue. This method is especially useful when the individual responses within a group, to some experimental perturbation, are highly variable; where functional impairment in the most sensitive individuals of the group may be explained by altered stereological values.

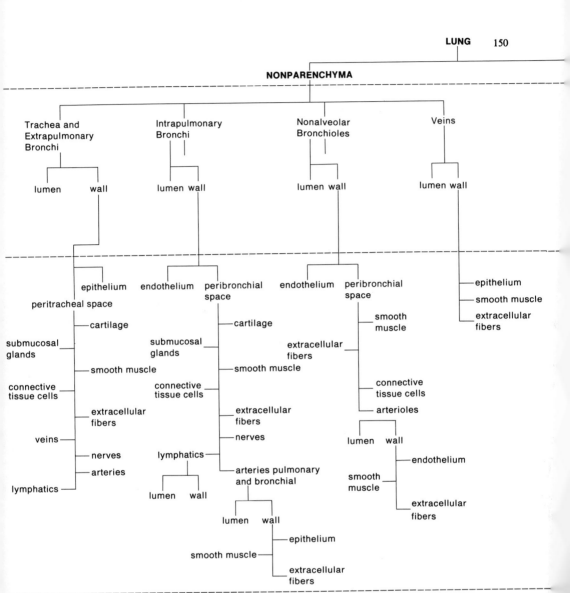

Fig. 12.1. Flow chart of the structural organization of a typical mammalian lung.

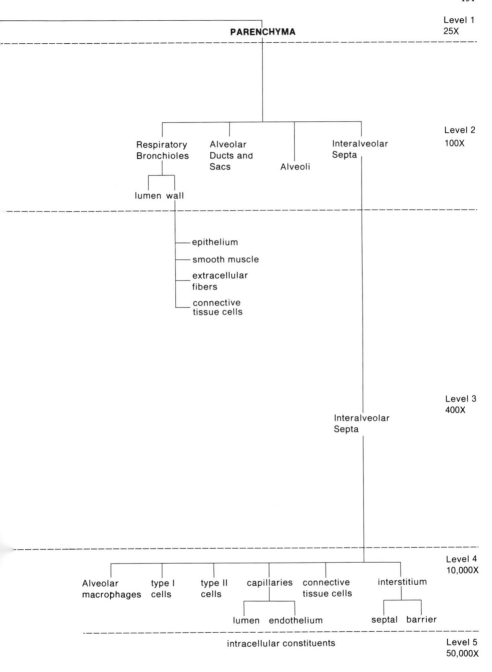

In each of the following examples, we attempted to relate our stereological measurements back to the primary function of the organ or to its pathology. We are confident that these may serve as models to other organ systems and to other pathologic conditions.

The Lung*

In our approach to the lung, we encountered some initial conceptual problems of precise physiologic fixation. The lung is suspended in the chest by its hilar attachments and a gradient of pleural pressures (*West* 1972); hence, whenever the thoracic cavity is entered surgically, the physiologic state of the lung is altered. However, if only the maximum capacity of lung structure is to be evaluated, the lung can be fixed at a pressure that approximates total lung capacity (TLC). This pressure can be found by examining saline pressure volume curves in the experimental animal of choice.

It is always desirable to estimate the amount of fixation shrinkage or swelling in order to relate stereological measurements of fixed tissue to fresh tissue. Previously, we have utilized the following formula (*Hyde et al.* 1978) after *Thurlbeck* (1967):

$$f = \left(\frac{\text{TLC} + W_L}{V_L}\right)^{1/3} \qquad (12.1)$$

to estimate fixation shrinkage or swelling (f) using the relationship of TLC plus lung weight (W_L) to the displacing volume of the lung (V_L) fixed at TLC.

A more accurate and recent method, using radiographic estimates of lung volume in the chest, has been presented for the human lung (*Pierce et al.* 1979; *Thurlbeck* 1979). A method we have used recently is to inflate individual lung lobes postmortem with air at 30 cm of H_2O pressure, measure the displaced volume, fix the lung lobe via intratracheal instillation of fixative at 30 cm of H_2O pressure after the lung is allowed to collapse to minimal volume, and then measure the displacing volume of the fixed lobe. Using this method, f^3 is equal to the ratio of the air-inflated to fixative-inflated volume. This latter method has proved to be reproducible in smaller mammals (i.e., rodents to small carnivores). Since we routinely use glutaraldehyde fixatives, the speed with which they are instilled is an important factor in reproducibly fixing lungs to TLC (*Hayatdavoudi et al.* 1980). To insure maximum flow at 30 cm of fixative pressure, we use perfusion tubing greater in diameter than the trachea.

---

* The authors would like to thank Mr. E. Reus, Ms. N. Tyler, Dr. W. Tyler and Dr. C. Plopper for writing the computer programs and assisting with this section.

Figure 12.1 presents a flow chart of the structural organization of a typical mammalian lung. Our initial approach is to slice lung lobes into 5-mm-thick slabs in a frontal plane (fig. 12.2). Subsequently, the slices are evaluated for volume densities of nonparenchyma and parenchyma, using an orthogonal lattice grid in a stereomicroscope at a magnification of 25 ×, level 1 (fig. 12.3).

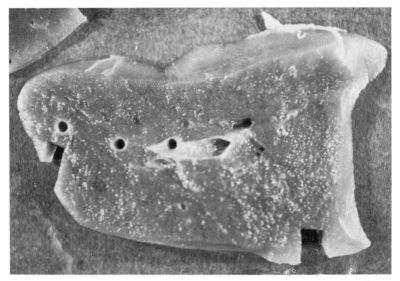

Fig. 12.2. A 5-mm-thick slab from a frontal plane of a right caudal lung lobe of a mammal. Large airways and vessels are easily distinguished from the surrounding parenchyma.

Subsequent sampling from the sliced lung lobes is dependent on the focus of the investigation. In an ongoing study of the pulmonary effects of chronic, low-level diesel exhaust exposure in adult cats, lesions are anticipated to focus on the nonalveolar and respiratory bronchioles. Also, some generalized fibroplasia is expected in interalveolar septa. In anticipation of these lesions, the following sampling scheme was used for each lobe.

Stratified random sampling was applied to lung slices. All lung slices were numbered consecutively beginning with the most dorsal slice. All odd numbered slices were used for random selection of segmental/subsegmental bronchi and nonalveolar/respiratory bronchioles. This was accomplished by bisecting these specific airways with a new clean razor blade which gives

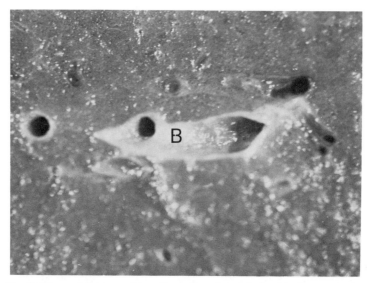

Fig. 12.3. Lung slabs are used to estimate the volumetric density of nonparenchymal and parenchymal components using a stereomicroscope at low magnification (10 to 25 ×). A bronchus (*B*) and a pulmonary vessel are visible in the field.

us two complementary halves per lung slice (fig. 12.4a). Subsequently, one of these halves can be used for estimates of wall components in airways of comparable size at level 3 (fig. 12.1). The other complementary half is routinely used for observation of airway surface morphology by scanning electron microscopy (fig. 12.4b). The surface morphology of nonalveolar bronchioles was essential in estimating the extent of a nonciliated bronchiolar cell hyperplasia in response to chronic air pollution exposure (*Hyde et al.* 1978).

Recently, *Plopper et al.* (1981) used an airway dissection technique to estimate the distribution of cell types in airways of identical location in different lungs. In this method the airway identification techniques developed by *Phalen et al.* (1978) are employed, using airway casts. The application of this technique to the sampling requirements of stereology could prove very useful in approaching a quantitative estimate of epithelial and wall components of airways of the same generation and/or diameter. Such information would be very useful in the investigation of chronic obstructive lung disease.

Stratified random sampling was also applied to even numbered slices in which blocks 5 mm × 1 cm^2 are selected at low primary magnification

Applications of Stereological Methods to Various Organs 155

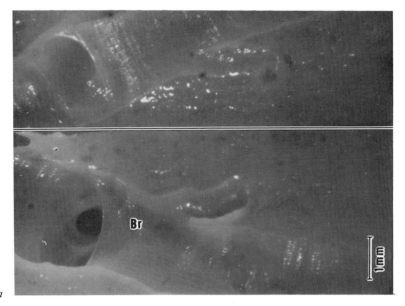

Fig. 12.4. Airways are bisected with a new, clean razor blade to obtain complementary halves. (*a*) A segmental bronchus (*Br*) gives rise to a very distal airway. (*b*) Subsequent examination of one complementary half enables easy identification of branching distal airways. In this case a final order of bronchus (*Br*) bifurcates into two terminal bronchioles (*Tb*). (Bar *a* = 1 mm and *b* = 100 μm).

(about 10×) using a stereomicroscope and an orthogonal lattice grid (fig. 12.2). Random numbers are generated, using a calculator, where the upper left corner of the block serves as the origin of the coordinate system. Following this procedure, a stratified sample of blocks is obtained between lobes and within lobes.

Samples used to determine interalveolar septal components at level 4 (fig. 12.1) are selected by cutting a 1-mm slice from each block taken from even-numbered slices of lung lobes. These 1-mm slices are about 5 mm deep and can be cut into 1-mm^3 blocks for postfixation with osmium tetroxide and plastic embedding for transmission electron microscopy. However, before any block (large or small) is postfixed and embedded, its cut surface is traced, using a drawing attachment on a stereomicroscope. Characteristic features of a block face, such as airways or vessels, are also drawn so that the face of the block sketched can be embedded toward the microtome knife. Subsequent to tissue sectioning, staining, and mounting on glass slides (for large blocks) or grids (for small blocks), the sectioned tissue face can be resketched at the same magnification and a tissue processing factor (p) calculated by the following formula:

$$p = \left(\frac{\text{fixed tissue area}}{\text{sectioned tissue area}}\right)^{1/2} \tag{12.2}$$

In all of our stereological calculations, fixation (f) and processing (p) correction factors are used. One must be aware that f and p are used for linear corrections, f^2 and p^2 for surface corrections and f^3 and p^3 for volume corrections.

Large plastic sections, 1 μm thick, can be observed using a light microscope at a magnification of 100× and an orthogonal lattice grid. At this magnification (level 2, fig. 12.1) the volumetric densities (V_V) of nonparenchymal and parenchymal components can also be estimated by stratified random sampling of the sections. One micron sections and light microscopy are used to estimate V_V of parenchyma and nonparenchyma of small lungs; with large lungs, gross lung slices and a stereomicroscope are used.

Similarly, at this same magnification the arithmetic mean thickness (τ, a volume-to-surface relationship of a tissue sheet) of walls of bronchi, vessels, or nonalveolar bronchioles can be determined from the following formula (after *Weibel and Knight* 1964):

$$\tau = \frac{d \cdot P_t}{4 \cdot P_i} \tag{12.3}$$

where d is the unit distance between test points for regular pattern grids, P_t the number of test points falling on tissue (wall in this case), and P_i is the number of intersections with the tissue sheet.

For parenchymal components, some general rules of recognition must be formulated. The following architectural differences are used to distinguish the various components of the pulmonary acinus: respiratory bronchioles have alveolar outpocketings and continue distally until the width of the cuboidal epithelium separating alveoli is less than half the width of adjacent alveoli; alveolar ducts and sacs are airspaces with three or more centrally projecting septa and/or distal continuations of respiratory bronchioles; and alveoli are squamous epithelium lined airspaces where interalveolar septa must project at least half the length of adjacent alveolar diameters to separate two alveoli. Alveolar surface area can be estimated at level 2, 3 or 4 (fig. 12.1); however, since it is influenced by resolution (*Keller et al.* 1976), we recommend that S_V be estimated at level 3 on 1 μm sections.

Within level 3 the usual differentiation is epithelium and the rest of the wall components (fig. 12.5). Individual cells of epithelial and connective

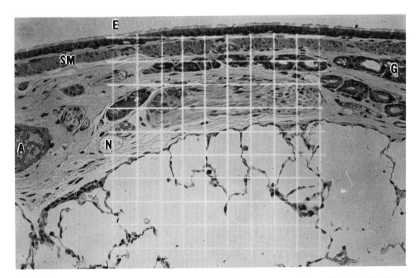

Fig. 12.5. The wall of a segmental bronchus which demonstrates bronchial glands (*G*), a bronchial artery (*A*), nerve (*N*) smooth muscle (*SM*), bronchial epithelium (*E*) and adjacent parenchyma.

tissue can be better differentiated within level 4. However, a reasonable estimate of cell types and volume estimates of cells versus other components can be obtained at level 3.

Components of interalveolar septa can be easily differentiated at level 4, 10,000 × magnification. One region of specific interest in the interalveolar septum is the alveolocapillary membrane, formed by type I epithelial cells, barrier interstitium, and capillary endothelium (fig. 12.6). Arithmetic mean

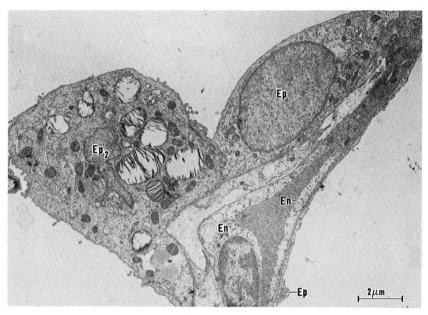

Fig. 12.6. An electron micrograph of an interalveolar septum of a mammal shows the components of the alveolocapillary membrane which is composed of the attenuated type 1 epithelial cell (Ep), capillary endothelium (En) and their fused basal laminae. Nuclei of type 1 and type 2 epithelial (Ep_2) cells as well as that of a capillary endothelial cell are also visible. (Micrograph provided through the courtesy of Dr. W.S. Tyler.)

thicknesses of all portions of the alveolocapillary membrane can be estimated by formula 12.3. *Weibel and Knight* (1964) have also introduced a method for determining the harmonic mean sheet thickness, a measure of the minimal thickness of a tissue sheet, by the following formula:

$$\frac{1}{\tau h} = \frac{3}{2} \cdot \frac{1}{\ell_h} = (1.5) \frac{\sum\limits_{i=1}^{n} fi \cdot \frac{1}{\ell_i}}{\sum\limits_{i=1}^{n} fi} \tag{12.4}$$

where ℓ are individual measures of sheet thickness in a variety of size classes, i, which are multiplied by their frequency f. As is evident, this measurement is influenced much more by minimal than by maximal values. Hence, it is not unexpected that *Weibel* (1970) used this measurement in deriving a diffusion capacity model of the lung. In practice, ℓ is measured using an orthogonal lattice grid and a ruler (fig. 6.11). Random measurement of ℓ is accomplished by measuring intercepts along the lattice grid lines. *Weibel* (1970) recommends the use of a logarithmic scale ruler since small values must be precisely measured. Our experience confirms the validity of this recommendation.

The components of the intracellular constituents are not presented in figure 12.1; however, these are provided in table 12.1 with abbreviations for all pulmonary constituents encountered.

Table 12.1 Symbols for pulmonary constituents

Levels 1 and 2
NP: nonparenchyma
P: parenchyma
T: trachea and extrapulmonary bronchi
B: intrapulmonary bronchi
NB: nonrespiratory bronchiole
V: vein
LU: lumen
W: wall
rb: respiratory bronchiole
ad: alveolar duct and sac
a: alveolus
t: interalveolar septal tissue

Level 3
E: epithelium
C: cartilage
G: submucosal glands
M: smooth muscle
CT connective tissue cells
EF: extracellular fibers
N: nerves
AB: bronchial arteries
AP: pulmonary arteries
L: lymphatics
lu: lumen
w: wall
AT: arteriole
EN: endothelium

Level 4
cl: capillary lumen

continued

(Table 12.1 continued)

cen: capillary endothelium
si: septal interstitium
bi: barrier interstitium
ct: connective tissue cells
ep1: epithelial type I cell
ep2: epithelial type II cell
am: alveolar macrophage

Level 5, Intracellular constituents
n: nucleus
cy: cytoplasm
mt: mitochondria
ly: lysosome
ll: lamellar body
gr: granules
rer: rough endoplasmic reticulum
ser: smooth endoplasmic reticulum
go: Golgi apparatus
nl: neutral lipid
v: vesicles and vacuoles
pm: plasma membrane
mv: microvillus plasma membrane
ci: cilium
mto: mitochondrial outer membrane
mti: mitochondrial inner (cristal) membrane

Within the frame of reference of figure 12.1 we will show some examples of the estimation of various parameters in a typical pulmonary investigation (*Hyde et al.* 1978). Lung volumes were estimated using weight displacement (fig. 3.3) and an f value calculated using formula 11.1 for 12 normal beagle dogs. An average of 12 lung slices were obtained from the right cranial, middle, and caudal lobes (fig. 12.2). Volumetric densities of 11.2% and 88.8% were obtained for nonparenchymal and parenchymal components, respectively, using a stereomicroscope and an orthogonal lattice grid. Since the focus of the investigation was on parenchyma, nonparenchymal components were not estimated. Stratified random blocks were used to estimate the volumetric densities of parenchymal components using the light microscope at level 2 (fig. 12.7). The following volumetric densities were estimated after a minor correction for the Holmes effect (*Weibel* 1963): for parenchymal components, respiratory bronchioles (wall + lumen) = 4.81%; alveolar ducts and sacs, 24.86%; alveoli, 57.31%; and interalveolar septal tissue, 14.02%. Hence, in order to obtain the volumetric density of alveoli for the entire

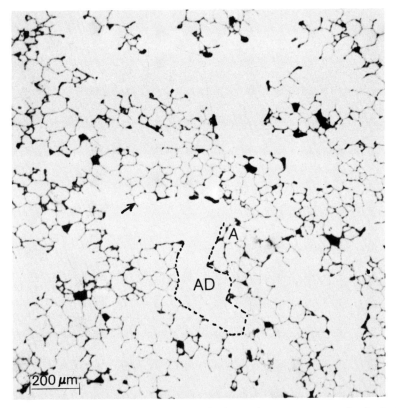

Fig. 12.7. A low magnification light micrograph of parenchyma components, alveolar ducts (*AD*), alveoli (*A*) and interalveolar septa (arrow). A dashed line separates an alveolar duct lumen from alveoli.

lung, it would be necessary to standardize it to a new reference volume by the following formula:

$$V_{va} = (V_{va}/V_{vp})(V_{vp}/V_L), \qquad (12.5)$$

which is simply the volumetric density of alveoli (V_{va}) in reference volume parenchyma (V_{vp}) multiplied by the volumetric density of parenchyma in the reference volume of the lung (V_L). In this case, it would be

$$V_{va} = (57.31\%)(88.8\%) = 50.89\%, \qquad (12.6)$$

which tells us that at TLC about half of the lung volume resides in the alveolar lumen. We can use formula 3.4 and an intersection count of interalveolar septa to estimate the surface area of interalveolar septa per unit volume of parenchyma (S_V). For 12 dogs, we estimated that S_V had a mean of 7.2042 × 10^4 mm^2/mm^3.

Let us proceed to level 4 using thin sections (50–80 nm) and transmission electron microscopy to investigate the alveolocapillary membrane. Using formulas 12.3 and 12.4, the arithmetic mean thickness of the alveolocapillary membrane is 1.7191 μm and the harmonic mean thickness is 0.4939 μm. It is also possible to estimate the contribution of each membrane component to either of these two thickness measurements.

Estimates of volumetric densities could be continued to include cell types. For example, a volumetric density of type I and type II epithelial cells in reference volume parenchyma is estimated to be 1.14% and 0.54%, respectively. These densities can also be estimated in the reference volume of interalveolar septa. Further, using the methods of three-dimensional reconstruction and estimation of mean caliper diameter, presented in Chapter 11, the number of each cell type per unit volume of interalveolar septal tissue can be estimated using formula 5.3 and mean cell volume using formula 5.36. These measurements are of obvious importance in characterizing the potential of cellular hyperplasia and hypertrophy, respectively. Investigations at level 5 follow from questions posed at level 4. For example, we might want to characterize the nature of a cellular hypertrophy, especially if there is a subjective impression of an increase in some subcellular component.

There is an almost unlimited number of features that potentially would be desirable to evaluate stereologically, such as fibroplasia and smooth muscle hyperplasia in walls of airways and arteries; however, a few examples of estimating components in parenchyma have been given as an illustration of a stepwise stereological approach to the mammalian lung.

A few FORTRAN IV computer programs that are currently used in our laboratories are presented in Appendix 2 as an aid to those readers who would like a more efficient method of data handling than data sheets and a calculator. These programs have been extremely useful for the collection, storage and processing of stereological data.

*Cardiac Tissue**

The approach to the heart is almost identical to that of the lung. Precise physiologic fixation of the heart is desirable and a flow chart is used for the

---

* The authors would like to thank Mr. J. Inderbitzen and Dr. D. Buss for writing the computer programs and assisting with this section.

structural organization of cardiac tissue (fig. 12.8). It is possible to fix the heart in one physiologic state—that of diastolic arrest and maximal coronary vasodilatation. This state of fixation is accomplished by anesthetizing an animal with intravenous sodium pentobarbital and administering 150 units of heparin per kilogram body weight. A thoracotomy is performed and the heart removed. The unfixed volume of the heart is estimated by weight displacement (fig. 3.3). The heart is subsequently connected to an isolated heart pump via the aortic root. Diastolic arrest and maximal coronary vasodilatation are accomplished by perfusing procainamide hydrochloride and adenosine in unoxygenated Tyrode's solution, respectively. A perfusion pressure of 100 mm Hg is used for this initial perfusion with a vented left ventricle, until the coronary vasculature is relatively free of blood (about 3 minutes). Then the perfusion is switched to a fixative solution of 3% glutaraldehyde in 0.2 M sodium cacodylate buffer at a pH of 7.3 for 15 minutes. This method insures fixation of the heart at a minimal, reproducible level of intra- and extravascular resistance. The fixed volume of the heart is then determined. The fixation factor, f, is calculated from the cube root of the fresh-to-fixed heart volume ratio. If postmortem angiograms of the coronary vasculature are desired, the aortic root can be recannulated and perfused with a warm (38°C) barium-gelatin solution at 140 mmHg. The barium-gelatin solution is then solidified by placing the heart in cold fixative.

Tissue selection from the fixed heart depends on the nature of the investigation. In any case, the atria and ventricles are usually separated by dissection and weighed. The right ventricular free wall can then be separated, leaving the interventricular septum with the left ventricle (fig. 12.9). This method allows reproducible weight, volume, and morphometric measurements among a variety of hearts.

As an example of sample selection and stereological treatment of cardiac tissue, we will describe one method for estimating stereological values of the coronary microvasculature. An apex to base measurement is taken of the dissected left ventricle (LV) and interventricular septum. A distance of 14% of the LV axis from the atrioventricular groove is used to cut a transverse section through the LV, which is 19% of the LV axis thickness (fig. 12.9). A protractor is used to measure borders of a 40° wedge from the regions of the anterior (APM) and posterior papillary (PPM) muscles, although the papillary muscles per se are excluded from the sample. Full wall thickness samples of wedges from APM and PPM are divided into subendocardial (ENDO), intermediate (INT), and subepicardial (EPI) layers, with eight blocks randomly selected from each of the six primary sample sites. Half of the blocks are cut perpendicularly to myofiber orientation, while the remainder are cut parallel to the myofiber direction. As with lung tissue,

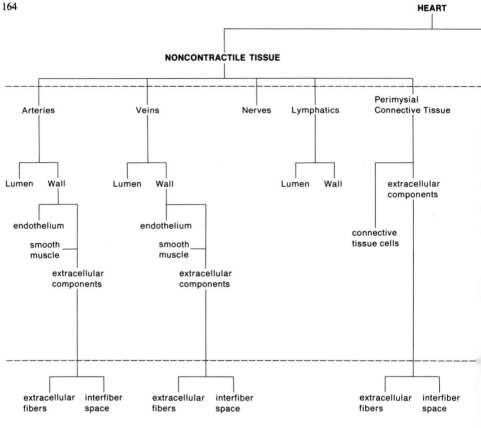

Fig. 12.8. Flow chart of the structural organization of a typical mammalian heart.

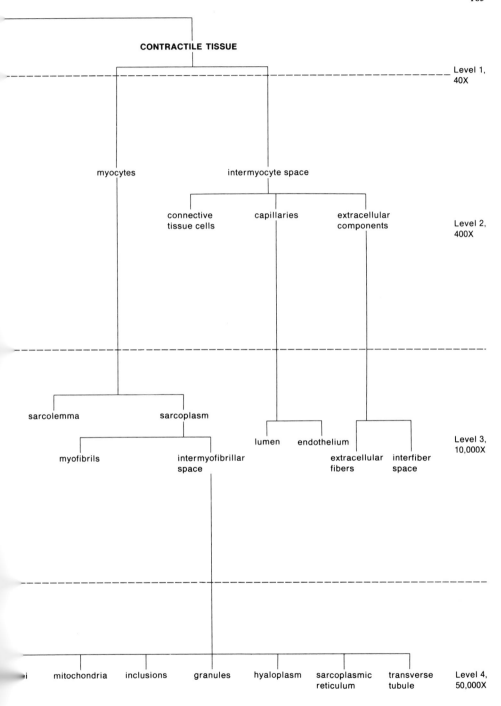

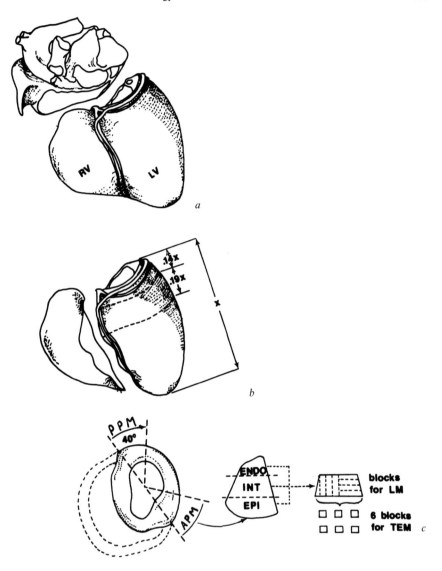

Fig. 12.9. Tissue selection from the heart. The atrial cap and right ventricular free wall are removed (*a*). The apex to base measurement (*X*) of the left ventricle (*LV*) is used to select a slice of LV wall, 19% of *X* (*b*). Two 40° wedges of tissue are cut from the anterior (APM) and posterior (PPM) papillary muscle regions (*c*). From each of these wedges, 3 transmural layers (subendocardial (ENDO), intermediate (INT) and subepicardial (EPI)) are selected. From each of these transmural regions, tissue blocks are cut for light and transmission electron microscopy.

these blocks are about 5 mm thick with a face of 10 mm². Slices 1 mm thick are cut from each block and cut into 1 mm³ blocks for transmission electron microscopy. All blocks are processed identically to that described for lung tissue, and a tissue processing factor is calculated by formula 12.2.

A light microscope with an orthogonal lattice grid (fig. 3.1) is used to estimate the volumetric densities of the primary noncontractile components and contractile tissue (fig. 12.10). Usual stereological values for canine left ventricle vary with species and regions; however, mean values over the LV are about 3% for noncontractile and 97% for contractile tissue. Volumetric densities of noncontractile tissue components can be very informative in estimating regional development of collateral coronary arteries.

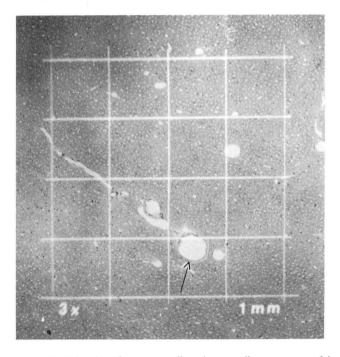

Fig. 12.10. Estimation of noncontractile and contractile components of the subendocardial region of the left ventricle using low magnification light microscopy. A large coronary artery is marked by an arrow.

Using a method similar to that for pulmonary arteries, the arithmetic mean thickness of coronary arterial walls can be estimated using formula 12.3. The volumetric components of arterial walls can also be estimated within level 2. Also, within level 2, the components of contractile tissue can be adequately estimated. Since levels 3 and 4 require electron microscopy and thin sections, stereological measurements within level 2 are much more economic in nature. If our investigation is not concerned with intracellular components, then level 2 measurements are usually adequate. For example, the mean LV myocyte and intermyocyte space volumetric densities are about 98% and 2%, respectively. Within the total myocyte volume it is possible to estimate the numerical density and mean volume of myocytes by three-dimensional reconstruction and estimation of mean caliper diameter (Chapter 11). *Loud et al.* (1978) have recently applied the method of numerical profile density regression on section thickness to obtain the numerical density of myocytes and their mean volume (formula 5.36).

Anversa et al. (1978) have used these methods to examine myocardial hypertrophy resulting from experimentally induced renal hypertension. With LV weight increased 30% and LV wall thickness increased 42% in a hypertensive group of rats, they found no change in the total number of myocytes; whereas, myocyte hypertrophy, as estimated from mean cell volume, was 76% greater in the subepicardial than subendocardial region. They carried the investigation into levels 3 and 4 (fig. 12.8) to estimate alterations of the subcellular components of myocytes, which they correlated with protein synthesis in a parallel investigation to better explore the mechanism of myocyte hypertrophy. Their approach demonstrated that within hypertrophied myocytes, myofibrils, sarcoplasmic reticulum and transverse tubules exceeded mean cell growth, whereas mitochondria, nuclei and other cytoplasmic components lagged behind.

Quantitative estimates of the coronary microvasculature are not as straightforward. Recent investigations of normal rat LV myocardium provide different conclusions relative to capillary profile density. *Gerdes et al.* (1979) reported a capillary profile density that significantly favored the epicardium, while *Rakusan et al.* (1980) found no difference in transmural estimates. Since capillary luminal surface density per volume of LV myocardium in the rat has been reported not to have a transmural difference (*Anversa et al.* 1978), we believe the reported transmural differences in capillary profile density may be the result of sampling problems of a partially oriented component of the intermyocyte space.

For all oriented structures (see Chapter 8), we recommend the use of a random or isotropic test grid. The Merz grid (1967) is a linear grid constructed from semicircles that one of us (DMH) uses for all oriented structures (fig. 3.1i), while the other (HE) uses an orthogonal grid, which can be rotated to abolish microscopic anisotropy. Capillary surface density determinations of myocardium can be confidently estimated using a Merz grid and block selection that maximizes all orientations (fig. 12.9). However, the expression of capillary profiles per area is highly susceptible to orientation. From scanning electron micrographs of coronary capillaries (fig. 12.11), we get a sub-

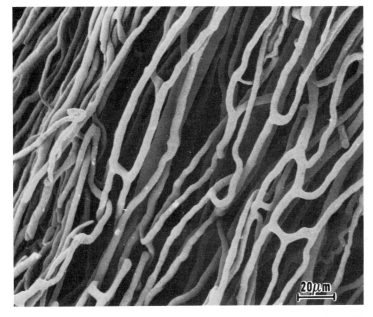

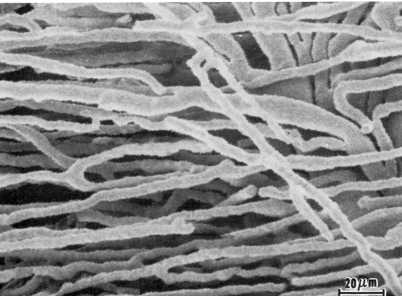

Fig. 12.11. Scanning electron micrographs of low viscosity resin-injected microvasculature of (a) the subendocardial layer and (b) the subepicardial layer of the posterior papillary region of the canine left ventricle.

jective impression of differences in the amount of capillary orientation in subepicardial and subendocardial regions. To estimate the potential transmural orientation difference of coronary capillaries, formula 8.3 was used and the results are recorded in table 12.2. It is evident from the data that a transmural difference exists in capillary orientation. The lower degree of orientation of subendocardial capillaries is derived from the greater number of capillary anastomoses, as evidenced by capillary casts (fig. 12.11), and the greater degree of three-dimensional departure of capillary sheets from any preferred orientation (fig. 12.12).

Table 12.2 Degree of capillary orientation

	APM (%)	PPM (%)
ENDO	29	25
INT	19	31
EPI	39	42

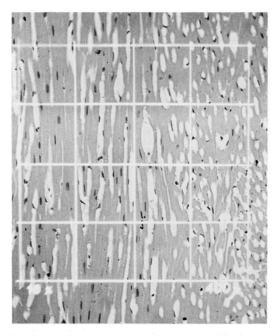

Fig. 12.12. A light micrograph of the subendocardial layer of the posterior papillary region of a canine left ventricle. Note the degree of divergence of myofibers and capillaries from any preferred orientation.

Both types of orientation can be estimated by axial ratio measurements of capillary profiles and used to standardize capillary profile densities by formula 6.8. Figure 12.13 shows the axial ratio measurement of a capillary profile. Table 12.3 compares the percentages for axial ratios of profiles of

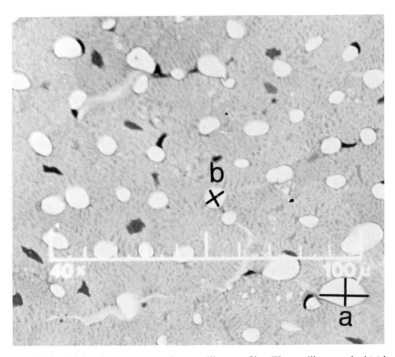

Fig. 12.13. Axial ratios are measured on capillary profiles. The capillary marked (*a*) has a ratio of about 2:1 while the one marked (*b*) has a ratio of about 1:1.

Table 12.3 Comparative percentages for axial ratios of coronary capillary and right circular cylinder profiles and mixed spaghetti embedded in gelatin

Axial ratio	Theoretical right circular cylinders (%)	Coronary capillaries (%)	Mixed spaghetti[a] (%)
1.0–2.0	86.6	80.4	83.3
2.1–4.0	10.2	16.1	13.6
4.1–8.0	2.4	3.5	2.5
8.0–∞	0.8	0.0	0.6

[a] Sectioned in 3 orthogonal planes.

coronary capillaries to those of right circular cylinders in random distribution and to mixed spaghetti embedded in gelatin and sectioned in three orthogonal planes.

It is evident that coronary capillaries do not vary much from a theoretical distribution of *straight* right circular cylinders or an actual distribution of right circular cylinders with random bending as mimicked by boiled, embedded and sectioned spaghetti. The difference of the distribution of profiles of coronary capillary axial ratios from the other two is the increased numbers of ratios from 2.1–4.0, which we believe is attributable to the anastomotic branches between capillaries.

The results of capillary profile density ($\#/\mathrm{mm}^2$) in six regions of 8 normal dog LV corrected by formula 6.8 showed no significant regional differences. Center-to-center intercapillary distance (ICD) can be calculated, assuming a hexagonal array of capillaries, by the formula:

$$\mathrm{ICD} = \left(\frac{2}{\sqrt{3}} \cdot \frac{1}{N_{AC}}\right)^{1/2} = \frac{1.0746}{N_{AC}} \qquad (12.7)$$

where N_{AC} is the numerical profile density of capillaries. The minimum diffusion distance (DD) is easily calculated by subtracting the mean capillary diameter ($\bar{D}_c$) from ICD divided by 2, or represented as a formula:

$$\mathrm{DD} = \frac{\mathrm{ICD} - \bar{D}_c}{2}. \qquad (12.8)$$

Mean capillary diameters are estimated from axial ratios of capillary profiles of 1 to 1 (fig. 12.13). Intercapillary distance, diffusion distance, mean capillary diameter, and capillary volumetric density are all useful descriptors of the coronary microvasculature.

Using thin sections and transmission electron microscopy, it is possible to further investigate the capillary endothelium at level 3 (fig. 12.14) by estimating the surface density per volume of contractile myocardium, the mean arithmetic and the harmonic mean thicknesses of the capillary endothelial wall. Results of these measurements in 8 dogs showed no significant regional differences.

Table 12.4 is a list of abbreviations of cardiac constituents usually encountered in an investigation. These abbreviations are used in Appendix 2 in a program currently used in our laboratory. The program utilizes profile axial ratio measurements, point-count and linear intersection methods described in this chapter. We hope these programs will save our readers time in executing a stereological investigation of the heart.

Applications of Stereological Methods to Various Organs 173

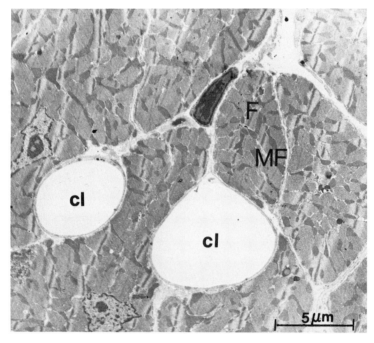

Fig. 12.14. An electron micrograph of canine left ventricular myocardium at level 3 magnification. Two capillary lumina (*cl*), a fibroblast nucleus (*F*) and myofibrils (*MF*) are clearly visible at this magnification.

Table 12.4 Symbols for myocardial constituents

Level 1
NC: noncontractile tissue
C: contractile tissue
A: artery
V: vein
N: nerve
L: lymphatic
P: perimysial connective tissue

Level 2
lu: lumen
w: wall
e: endothelium
sm: smooth muscle
cc: extracellular components
ct: connective tissue cells

continued on following page

(continued from page 173)

m: myocytes
im: intermyocyte space
c: capillary

Level 3
ef: extracellular fibers
is: interfiber space
sl: sarcolemma
sp: sarcoplasm
mf: myofibrils
if: intermyofibrillar space
cl: capillary lumen
ccy: endothelial cytoplasm
en: endothelial nuclei

Level 4
mn: myocyte nuclei
mt: mitochondria
gr: granules
nl: neutral lipid
v: vesicles and vacuoles
sr: sarcoplasmic reticulum
t: transverse tubule
h: hyaloplasm

Colonic Adenoma

An adenoma of the large intestine is a neoplasm which, according to conventional nomenclature, does not invade the submucosa (fig. 12.15), while a carcinoma is one that can invade through the muscularis mucosa and deeper.

Both of these lesions can be investigated successfully by stereological methods and clues to their behavior can be obtained. An adenoma is relatively easy to deal with since its confinement to the mucosa facilitates measurement, while a carcinoma, after invasion and metastatic spread, escapes from the territory of easy observation. Moreover, the structure of colonic adenomas exhibits a high degree of uniformity and remains a highly organized neoplasm.

The following short account demonstrates a chain of observations, through measurements and calculations, which possibly could lead to interesting pathophysiologic insights, and serves as an example of the potential usefulness of stereology. From the point of view of stereology, several questions can be asked and answered. Fundamentally, these are: (a) In what respect does the histologic architecture of an adenoma differ

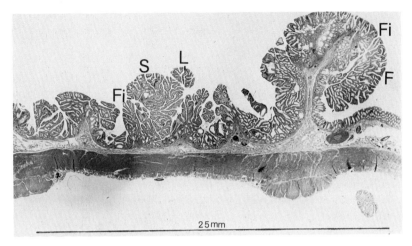

Fig. 12.15. An expanse of colonic adenoma, organized in one location into a pedunculated polyp. Normal mucosa at the far right. *F*: folium; *Fi*: fissure; *L*: lobe; *S*: sulcus.

from that of normal colonic mucosa and how could this altered structure change the function of an intestinal area? (b) How great is the growth factor? How much has the volume of the tissue increased? Compared with normal mucosa, how much is the epithelial surface area, which has arisen from a specific, circumscribed mucosal area, increased? How much has the number of epithelial cells increased? Is the increase in the number of cells related to the mitotic index or to some other factor? We will repeat here the essential steps and results in a general manner.

Structure, Functional and Carcinogenetic Significance

Colonic adenomas are traditionally described as tubular or villous. The different identifications depend on the direction of sectioning. The law of dimensional reduction, however, shows that colonic adenomas are neither tubular nor villous but consist of broad, flat *folia* (*F*) with extended, equally flat *sulci* (*s*) between them (figs. 12.15 and 12.16). This structure is recognizable, in various directions of cutting, at high, medium and low magnification (figs. 12.15–17). Only commonsense reasoning, not mathematics, is needed to arrive at this conclusion. Yet, to readers who still doubt this structural identification (fig. 12.18), taken by scanning electron microscopy, will be convincing enough. This picture distinctly shows that in addition to narrow sulci, great fissures (Fi in fig. 12.15) divide an adenoma into lobes (*L* in fig.

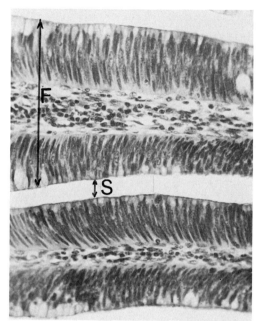

12.16

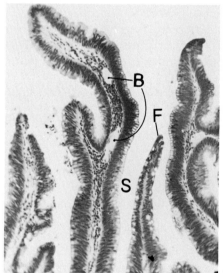

12.17

Fig. 12.16. Two adenomatous folia (*F*) with a sulcus (*S*) between. Note that the epithelium is highly pseudostratified.

Fig. 12.17. Section normal to the mucosa, showing adenomatous folia (*F*), sulci (*S*) and blood vessels (*B*).

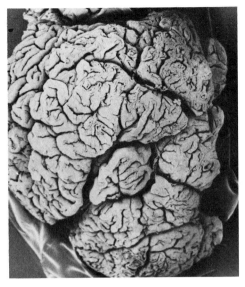

Fig. 12.18. An adenomatous polyp, dried and observed by scanning electronmicroscopy. Folia, sulci, fissures and lobes are evident.

12.15). The transformation of the tubular intestinal crypts into sulci and folia occurs by fusion of the lumina of neighboring crypts and by fusion of their epithelia (fig. 12.19). During this process, the original lamina propria (intertubular connective tissue) assumes the role of the core of the folia.

The original tubular crypts with their blind ends and one-way flow of mucus have been transformed by confluence into long and broad slits, which permit easy passage of ingesta. During the transformation, mucus is expelled from the epithelial cells, which then assume the morphologic character of indifferent or absorbing cells with variable numbers of microvilli. Thus two physiologic properties have been drastically changed. It is not only an increase in surface area, but also the creation of passable channels that permits greater absorption. This is even more facilitated by increased vasculature in the folia (fig. 12.17). The development of an adenoma *could* be a reaction to malabsorption in the small intestine. To determine whether this speculative thought is valid, many future clinicopathologic studies will be needed.

The easy passability of the adenomatous fissures and sulci may also facilitate contact of potential carcinogenetic material with the epithelium. The above are matters of shape determination, sometimes considered unworthy of a quantitative microscopist. Nevertheless, they fall into the realm of stereology.

Growth can be defined as an increase in volume, an increase in surface area, and as an increase in the number of cells. The desire to quantify any

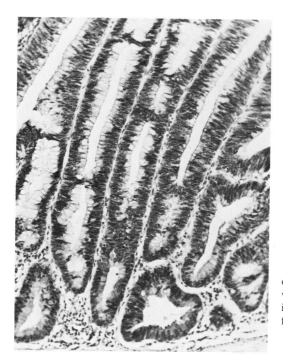

Fig. 12.19. Fusion of crypts and of epithelium whereby the colonic mucosa is converted to a system of folia and sulci.

aspect of growth presupposes knowledge of the quantitative aspect of the tissue of origin. In Chapter 6, when discussing the thickness of a lamina, we chose the colonic mucosa as a convenient example. Fortunately, the thickness of this layer can be determined from place to place. For oblique sections (fig. 12.20), the thickness T was found in formula 6.2 to be

$$T = W\sqrt{1 - \frac{1}{Q^2}} \qquad (12.9)$$

On the average, $\bar{T}$ was found to be 510 μm. The epithelial surface area could be found by determining the number of crypts per unit surface area and then, by calculating the average surface area of each, considering them as cylinders. But there is an easier way. Using random sections through the mucosa, the number P of intersection points of epithelium with test lines of known length (fig. 12.20, circles) are counted and the surface area per unit volume calculated by formula 3.4.

$$S_V = \frac{2P}{L} \qquad (12.10)$$

Applications of Stereological Methods to Various Organs

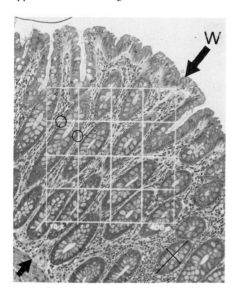

Fig. 12.20. Determination of normal mucosal thickness and of epithelial surface area per unit volume.

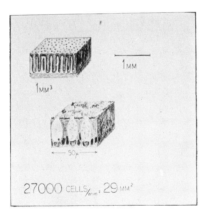 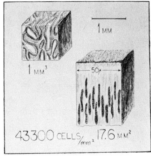

Fig. 12.21. (*Left*): Basic dimensional and numerical data for typical normal colonic mucosa. Above: 1 mm³ of mucosa. The entire white square represents the epithelial surface area within that tiny block of mucosa. (*Right*): The same information for 1.0 mm³ of a typical colonic adenoma.

The average epithelial surface area per cubic millimeter of normal colonic mucosa was found to be 29.4 mm^{-1} ± 3.5%. The number of epithelial cells in this same territory was determined by a count of nuclei in a test area A_T of tangential sections n/A_T (fig. 12.21). This number must be multiplied by the epithelial surface area per cubic millimeter to obtain the number of epithelial cells per square millimeter of normal epithelium, including the crypts: 27,600 cell/mm^2 ± 3.3%. Multiplying this number by the factor S/V, we obtain approximately 811,000 cells/mm^3 of normal mucosa (fig. 12.21, left).

As might be expected, the normal mucosa is not of constant thickness. For example, there are local depressions (fig. 12.22). In a section of a depressed

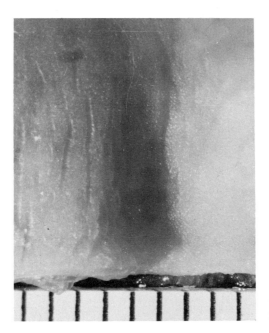

Fig. 12.22. A depression in the colonic mucosa (scale in millimeters).

area (fig. 12.23), it can be seen that this depression is due to attenuation of all layers of the interstitial wall. Such local singularities are exceptions. Since it is impossible to know the thickness, surface density, and cellular density of the mucosa of origin of a specific adenoma, it must be assumed that the average quantities existed. After the average normal mucosa is quantified, the same values are determined for specific adenomas. The most convenient examples are pedunculated polyps, such as is shown on the right

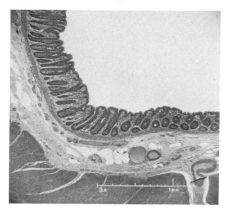

Fig. 12.23. Section through the edge of the depression shown in figure 12.22.

side of figure 12.15. These can be compared with average normal mucosa, which could have spanned the base of the pedicle.

Estimating the Volume of an Adenomatous Polyp

Only the adenomatous, thickened mucosa is measured, while the core made of submucosa is ignored (fig. 12.24); only an estimate is possible. Precise determination is not possible since the material is derived, from almost randomized sections, from the slide collection of the Surgical Pathology Laboratory. Assumptions must be made, the correctness of which cannot be verified. Yet the result is, however, an acceptable approximation to reality. The first assumption is: the polyp appeared circular when viewed from above, and the second assumption, the section passes through its central axis, perpendicular to the mucosal surface.

One draws an array of parallel lines of constant distance h through the image of the polyp parallel to the mucosa and measures the length of each line as long as it does not fall on stroma (submucosa) (fig. 12.24). This length d is assumed to be the diameter of the top surface of a circular slice through the polyp. Its area is

$$a = r^2\pi = \left(\frac{d}{2}\right)^2 \pi \qquad (12.11)$$

Then, one multiplies the sum of all the a's by h. The volume of the mucosal part of the polyp is

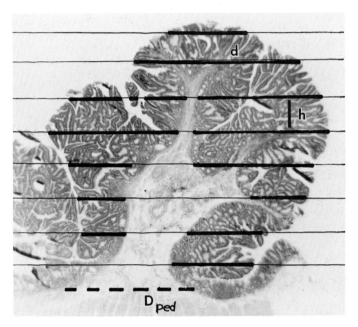

Fig. 12.24. Estimating the volume of a pedunculated colonic adenomatous polyp.

$$V_{\text{polyp}} = h \sum_{i=1}^{n} a_i. \tag{12.12}$$

This polyp is assumed to have originated from an area of normal mucosa coextensive with the base (dotted line D_{pedicle} in fig. 12.24). The volume of the mucosa at this location, before transformation, was approximately

$$V_{\text{pedicle}} = \left(\frac{D_{\text{pedicle}}}{2}\right)^2 \cdot \pi \cdot \bar{T} \tag{12.13}$$

($\bar{T}$ is the average mucosal thickness). $V_{\text{polyp}}/V_{\text{basal mucosa}}$ is the increase of volume.

Next we determine the epithelial area per cubic millimeter of adenomatous mucosa by an intersection count as shown in figure 12.25.

$$S_V = \frac{2P}{L} \tag{12.14}$$

Applications of Stereological Methods to Various Organs 183

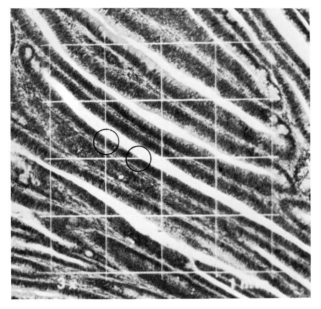

Fig. 12.25. Determination of epithelial surface area per unit volume by intersection count.

The total epithelial surface increase is:

$$\frac{S_{V_{polyp}} \cdot V_{polyp}}{S_{V_{normal\ mucosa}} \cdot V_{pedicle}} \qquad (12.15)$$

The number of cells per square millimeter of adenomatous epithelium is calculated by a nuclear count in tangential sections through the layer of nuclei. Since adenomatous epithelium is pseudostratified (fig. 12.16), one tangential section does not intercept all nuclei at one spot. Therefore, one must find, by inspection of many sections normal to the epithelium, a correction factor. This correction factor is the estimated number of nuclear levels in the adenomatous epithelium, and was found in our studies to be approximately 1.3.

Astonishingly, the epithelial surface area per cubic millimeter of adenoma is lower than that for normal mucosa (fig. 12.21, right). It has changed from 29 mm^{-1} to 17.6 mm^{-1}; i.e., by a factor of 0.606. This can be explained through the fusion of neighboring crypts. But the number of cells per square millimeter of epithelium has increased by a factor of 1.6. This greater cell

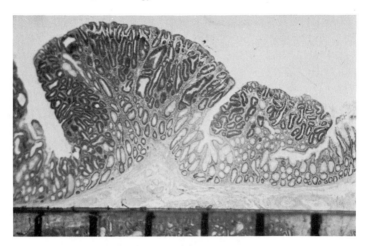

Fig. 12.26. Two very small adenomatous polyps. Scale in millimeters.

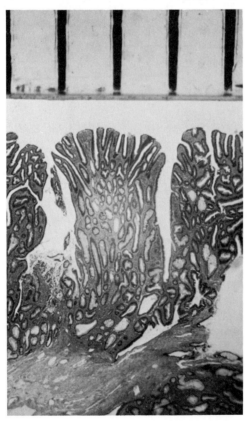

Fig. 12.27. A very small, cylindrical, almost sessile adenomatous polyp in the colon.

density is due to cellular cessation of mucus production, pseudostratified arrangement and crowding (fig. 12.19). The result is approximately the same number of cells per unit volume. Growth of the adenoma, hence, is expressed by an increase in volume due to an increase in cell number.

Figure 12.21 attempts to visualize these relationships. In the upper left hand corner of each of the illustrations a cubic millimeter of tissue is shown (figs. 12.26 and 12.27): left, normal mucosa; right, a block out of an adenoma. The large white area represents the epithelial area in that 1 mm³. At the bottom of each picture, a block of epithelium is drawn covering a surface of 1/400 mm².

Figures 12.28, 12.29 and 12.30 show four polyps of different sizes and shapes. Sessile polyps, i.e., those with broad bases, have less opportunity for relative growth. Pedunculated polyps exhibit greater relative enlargement. We have analyzed these quantitative aspects in eight polyps. For this brief

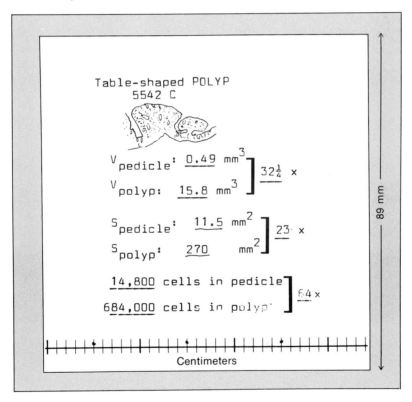

Fig. 12.28. The two polyps photographed in figure 12.26 with numerical data and scale. The white square equals the epithelial area of the left polyp.

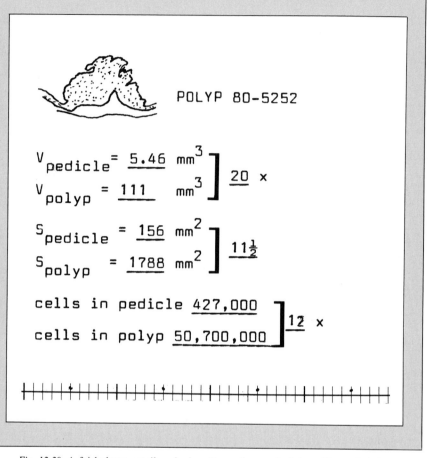

Fig. 12.29. A fairly large, sessile colonic polyp and numerical data. The scale is in millimeters. The white square equals the epithelial area of the polyp.

presentation, we have selected three of them; one sessile polyp with a volume increase of 20 × and a cell number increase of 12 × (fig. 12.29); one pedunculated polyp with a volume increase of 224 × and a cell number increase of 370 × (fig. 12.30), and another polyp, in absolute size the smallest of the three (fig. 12.28), with a volume increase of 32.52 × and a cell number increase of 64 ×. This is the same as that seen at the left in figure 12.26. Its relative volume increase is 32.25 ×. The increase in cell number is 64 ×. The white squares surrounding the drawings of the polyps represent the epithelial areas drawn to scale. In all of the eight cases, cell multiplication has been considerable (11–470 times).

It is customary to ascribe this growth, which we believe has been quantified numerically for the first time, to a lack of cells, since the mitotic index in these lesions is not essentially higher than in the normal mucosa. Lack of cell loss does not satisfactorily explain the increased growth. Rather, we

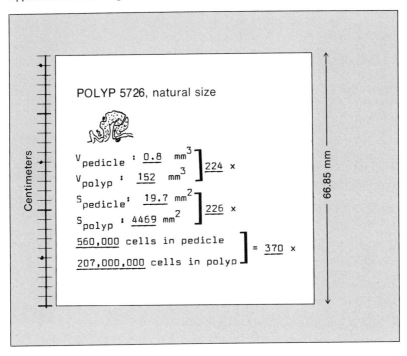

Fig. 12.30. The polyp shown at the right in figure 12.15 with numerical data. White square equals epithelial area in polyp.

ascribe it to multiple fragmentation of nuclei as described for colon carcinoma by *Elias and Fong* (1978). The same phenomenon can be observed in colonic adenomas (fig. 12.31).

Early in the fragmentation process the nuclear fragments called "nucleotesimals" are connected with the original nucleus by lamellar bridges through which nucleic acids could pass. Later, after repeated S phases, they seem to separate so that each can become the nucleus of a new neoplastic cell. It is a kind of amitotic or direct cell division. The phenomenon is capable of explaining the enormous growth of these neoplasms.

At this time it is difficult even for us to gain a visual, three-dimensional concept of a fragmenting nucleus. Serial section reconstruction would require about 60 consecutive thin sections, if cut parallel to the long axis of a nucleus. Even if a model were made, the parts would overlap and obscure each other. However, computer reconstruction, which produces rotating three-dimensional images (see Chapter 11), may become useful in their visualization.

Another problem requiring considerable stereological effort is the estimation of the percentage of nuclei fragmenting at any one time. To this difficulty are added the factors of circadian rhythmicity. Here is a new and

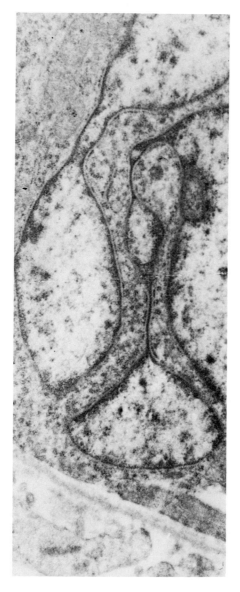

Fig. 12.31. Basal part of a nucleus of a colonic adenoma cell exhibiting fragmentation into nucleotesimals connected by lamellar bridges.

perhaps important challenge to future oncologists trained in stereology.

The above have been three examples of work in which stereology has played an important, if not the only role.

XIII. Potential Future Applications of Stereology

Introduction

The potentials of stereology have hardly been tapped. New opportunities for its useful application are all around us. The following are a few problems to which stereology could be applied. We offer proposals to our readers since we want the resources of stereology to be open to new vistas. Since we are both mortal, we cannot possibly complete all that we could if time were unlimited. But we can offer a few suggestions, and if we propose new avenues and, in a few cases, even describe possible methods, we still reserve the right to do this work by ourselves, if we find the time. We will open this list of stereological adventures with an item that appears humorous. The second item will be a view into hyperstereology, i.e., extrapolation from three- to four-dimensional space. After these two excursions, we will enumerate a few potential examples from the biomedical sciences.

The Swiss Cheese Problem

"Swiss cheese is made easily using a drill, by boring holes in any cheese that you will" (German nursery rhyme). If the method alluded to in this rhyme were actually used, the holes in Swiss cheese would be parallel cylinders. Slices through the cheese cut in random directions would show elliptic profiles with the value of Q, depending on the angle of sectioning. What results is Q distribution, as shown in figure 4.3a. Obviously, this is not the case since the holes in slices of Swiss cheese are always round, with Q near 1. We conclude that the nursery rhyme is false. The holes in three-dimensional space must be almost spherical; and the assumption that they are created by gas bubbles is plausible. If the slices were very thin, we would simply measure the diameters of the holes in the slices, classify them into size

---

* This section is reproduced in part from Elias, H.: Structure and rotation of barred and spiral galaxies. Interpreted by methods of hyperstereology (i.e. extrapolation from n- to $(n + 1)$-dimensional space; in Elias: Stereology, pp. 149–159 (Springer-Verlag, New York 1967).

classes, and construct a histogram. This could be matched with one of the curves in figure 5.10; or we would interpolate between two of the curves.

Fig. 13.1. Section through a slice of Swiss cheese.

Since cheese slices are, however, of finite thickness, a negative Holmes effect comes into play; negative because, in contrast to the effect originally described by *Holmes* (1921), not the particles but the matrix is opaque. Consequently, the smaller diameter of each profile will be visible, while the larger will remain hidden (fig. 13.1, right). Even the largest visible diameter must be smaller than the real diameter of a spherical bubble (fig. 13.1, left and middle).

This is all we can say about the Swiss cheese problem and leave the final solution to others. This may, indeed, become useful to the dairy industry.

Stereology of More-dimensional Spaces

As the visible spectrum encompasses only a very limited portion of electromagnetic waves, and just as it has been possible to extend its useful range into shorter and longer wavelengths by ultraviolet radiation, X-rays, and, at the other end, into infrared photography and radiotelescopy, so it is possible, by stereological methods—better called hyperstereology—to extend our knowledge of the universe by extrapolation from 3- to 4- or 5-dimensional hyperspaces (from S_3 to S_4 or S_5).

Elias in 1967 attempted to do this. The reader has only to recognize that the three dimensional space in which we live, and which we believe to comprehend, is only a limited portion of the world, limited in a manner similar to that of the range of the visible spectrum.

Every straight line perpendicular to all lines in our visible space points in the direction of a fourth dimension—as for example, any line perpendicular to the three edges that meet in an upper corner of a room (fig. 13.2)—although such a direction cannot be visualized.

n-Dimensional geometry is not a phantasy of the present authors; several textbooks have been written on the subject. One of them, by Forsythe,

Potential Future Applications of Stereology 191

Fig. 13.2. Upper corner of a room where 3 edges meet.

[1930] is concluded with the words: "if at any time in the future it should be found that our universe was curved, we would have to draw the necessary conclusion of the existence of a more extensive space." Six years later, Hubble [1936] found that it must indeed be curved, a concept now universally accepted.

Another 6 years later, Mayall and Aller [1942] found the astonishing fact that stars composing a spiral or barred nebula do not obey the third law of Kepler: their main disks are rigid. Even the structure of these galaxies, according to Kovaleski [1966], "has not been explained."

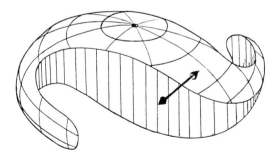

Fig. 13.3. Section through a hyper-spiral by a three-dimensional space.

The senior author [1967] attempted to provide an explanation, using sterological methods, simply by considering, with Hubble, the observable universe in which we live as an expanding hypersphere suspended in a a four-dimensional space.

The barrel and spiral nebulae—as the sky exhibits them—are to be considered sections of specific hypersolids cut by the expanding hemisphere that we call our universe. The hypersolids have cores shaped like the lower surface of winding staircases, the steps being straight for barred nebulae and parabolic for spiral nebulae. These hypersolids are oriented in the direction of the radius of curvature of the Hubble universe.

Rigorous mathematical methods have been used to develop a projection on S_3 of the assumed hypersolid (fig. 13.3). Two rotational phases of this section are shown in figure 13.4. (Interested readers should consult the original publication.)

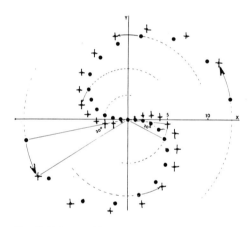

Fig. 13.4 Two positions of the core of a spiral galaxy at different times; i.e., once before (black dots) and once after S_c has expanded slightly (pluses). In S_3 the expansion and consequent change of position of the section manifests itself as a (pseudo) rotation.

As a model experiment, serial sections are made, by cinematography, through successive levels of an appropriately shaped object. While the double helicoid, three-dimensional object (fig. 13.5) is rigid, the projected image of its serial sections rotates on the screen.

Potential Future Applications of Stereology

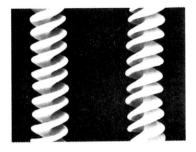

Fig. 13.5. A noodle called Spirella in two different positions.

Figure 13.6 shows several phases from that film. Figure 13.7 is a diagram of a hyperhelicoid cut at three different times by the universe, here shown as an arc of a circle. And, the outer portions of the rotating image have the same angular velocity as the inner, exactly as observed by Mayall and Aller [1942].

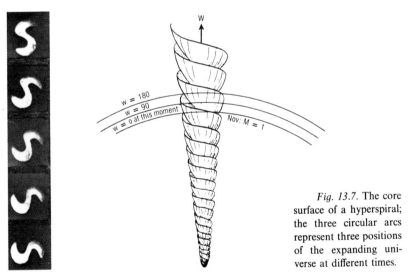

Fig. 13.7. The core surface of a hyperspiral; the three circular arcs represent three positions of the expanding universe at different times.

Fig. 13.6. (*Left*) Five frames from a serial section film: a rigid spirella noodle was embedded in paraffin and mounted in a sliding microtome, and the block surface photographed on a frame of film after each cut. When the film is projected, the succession of sectional levels through this rigid object appears as rotation (which in reality does not take place).

The importance of this four-dimensional model of spiral nebulae lies in the fact that it permits one to drop Gamow's "Big Bang" hypothesis as well as Bondi's [1960] assumption of the "continuous creation of natter."

Our four-dimensional model permits one to return to the assumption of a "steady state universe." Accordingly, the curved universe expands in S_4 like a wave front on a water surface S_2 in a bath tub. The expansion of successive wave fronts on a water surface may be due to excitation from three-dimensional space, such as a dripping faucet.

The model presented here permits the acceptance of a steady-state universe not necessitating a "big bang" nor a thinning out in the future of the universe. It permits a developmental theory as follows: compare the expanding universe with an expanding wave front. And in order to visualize the concept, let us step down to a two-dimensional plane, the surface of a body of water in a tub. Let drops from a leaking faucet fall down on the water surface, i.e., from a third dimension. A succession of waves expands from the point of impact, without changing the density of the water or of the air overlying it.

If we now return to four-dimensional space and to figure 13.7, the universe—presented by the circular arcs—can be visualized as taking in new matter from S_4, into which it expands. And, the assumption by *Lyttleton et al.*, as quoted by *Bondi* (1960), of "continuous creation of matter" becomes unnecessary. As on the water surface, energy for the expansion derives from an $(n + 1)^{st}$ dimension, in the case of the universe from a 5th dimension. The source of that energy cannot, at this time, be known.

As on a water surface, one wave front expands after the other. They all are concentric. And the older wave fronts are external to the younger ones. Thus, one can imagine that our universe is surrounded by older universes parallel to and concentric with it. We are entering a realm similar to *H.G. Wells' Time Machine* and *Manlike Gods*. We can also imagine that UFO's, if they exist, do not necessarily derive from far away planetary systems, but from the nearest expanding wave front, from an earth-like planet that is perhaps only a very short distance away from us, located in a direction of a fourth dimension, a direction that cannot be visualized.

These thoughts, as is all theory, are imaginary. Nevertheless, they may represent reality, just as atoms and elementary particles have been assumed to be realities long before they could be demonstrated.

Please do not take the above game too seriously. It is merely an exercise devised to demonstrate that stereology and hyperstereology are and may be capable of giving insight into realities beyond our present sensory perception.

Thomas Banschoff of Brown University has shown that mathematically derived *n*-dimensional structures that cannot be seen can be projected, always by pure, abstract mathematics, on a three-dimensional space. However, they can be projected on a two-dimensional computer-graphics screen, and thereby visualized (*Banschoff* and *Strauss* 1978).

The rigid rotation of galaxies and their barred or spiral structure is just one phenomenon not explicable on the ground of physics. It is not impossible that with time other things that defy an explanation based on three-dimensional physics might be understood on the basis of hypergeometry. Stereologists must keep their eyes and ears open to reports on inexplicable phenomena. Maybe they will be able to help their colleagues in other disciplines through their training in extrapolation.

The above four-dimensional theory arose by accident. Long ago, in the early days of stereology, it appeared desirable to understand and to corroborate the mathematical stereological methods, derived from theoretical reasoning, by model experiments to convince readers unfamiliar with geometric probability. Three-dimensional models of known structure were constructed, usually embedding solids of exactly known shape in random orientation within a continuous matrix. The entire assembly, an artificial "organ," after hardening, could be cut in random directions. On these sections, counts and measurements were made, and the mathematically arrived at formulas could be confirmed experimentally. For example, to confirm the mathematically predicted Q distribution of profiles of straight circular cylinders, *Hennig and Elias* (1963a) embedded wooden rods into plaster of Paris and after hardening, cut the mass in random directions with a carpenter's band saw. The axial ratios of the profiles were measured and tabulated. The expected distribution was confirmed. To confirm the theory of Q distribution of rotatory ellipsoids, hundreds of ellipsoids of predetermined shape were cast in colored plaster of Paris and embedded, after thorough mixing, in white plaster of Paris. Again, a band saw was used in the justified hope of confirming the theory. The same method was used by *Hennig and Elias* (1963b) to determine the Q distribution for profiles of triaxial ellipsoids, a problem that hitherto defied a mathematical solution. Other stereological problems were corroborated by *Elias* and coworkers, who used variously shaped noodle products embedded in blackened gelatin. When a new kind of noodle, "spirella," shaped like a double helicoid, came on the market, it was subjected to the same treatment and it was found, unexpectedly, that exact transverse sections through such a noddle looked like spiral nebulae. From the accidental game, the above cosmologic theory was developed. In 1955 *Elias* presented this theory to the section on Astronomy of the American Association for the Advancement of Science in Atlanta. A spokesman for the astronomers and cosmologists, present at that session, said that the approach might prove useful; none found a mistake.

And as contemporary astronomers have their hands, telescopes, and spaceprobes fully occupied, not running out of problems in this thin three-dimensional space of ours, so will the simple biologists now return to earth and try to apply stereology to problems of the biosphere.

Hematology

Much of quantitative hematology is performed by smears, centrifugation, and other methods through which blood cells are deformed. Using constant methods, currently used hematologic procedures are valid since they *compare* the blood of patients with the blood of healthy young people, which has been subjected to exactly the same artifactual treatment. Most of these hematologic methods do not, however, tell the truth about the living reality. Also, blood smears greatly distort blood cells. We will outline, now, some stereological and histologic procedures that might enable hematologists to compare their routine observations with the reality in the living body.

The first problem is to preserve blood in situ in a condition as faithful to life as possible. One of the problems to be investigated is to compare the hematocrit values, determined in the conventional manner, in various compartments of the intact organism.

Peripheral blood. Two specimens are collected from an animal. An anticoagulant is added to one (specimen [a]) and the specimen centrifuged. The other specimen without anticoagulant (specimen [b] is poured into a groove—perhaps Teflon coated—and allowed to clot. Once formed, the clot is removed, immersed in a suitable fixative, and cut at several levels. Data are obtained from specimen [b] by stereological analysis of semithin and thin sections, as shown in figure 13.8. The conventional procedure may have introduced artifacts which, at this time, are unknown. Centrifugation may have compressed the blood cells and may have resulted in an artificial *de*crease in cell volume and an *in*crease in supernatant volume. Stereology, will give more realistic values.

Blood in Various Vessels Including the Heart. Submerge a young adult nude mouse, under light anesthesia, in liquid nitrogen or freon. The entire animal will be frozen rapidly. Then immerse the frozen animal in a large volume of supercooled Carnoy's mixture at 20°C. Let the specimen, possibly under gentle agitation, sit at room temperature. The fixative will warm gradually and will thaw the specimen in thin layers, from outside inward. As specimen layers thaw, the fixative, which is known for its fast penetration, will follow the thawing process immediately. And, after 1 to 2 hours, the entire animal will be fixed evenly. Carnoy's mixture is a fast-acting fixative that permits excellent staining. As it fixes, it dehydrates. Since, for blood stereology, very thin cutting is required, embedding in glycol methacrylate should follow immediately after two changes into 95% alcohol. For illustration, we depend on a preparation, presently at our disposal, treated in a way different from that outlined above.

Figure 13.8 is from an interlobular artery in the kidney of a dog, fixed in glutaraldehyde, embedded in epon, cut at $\frac{1}{2}$ μm and stained with toluidin blue by Dr. Kurt Albertine. For stereological purposes, the method outlined above is likely to yield more faithful results. But the stereological procedure can be outlined using this photomicrograph. On the other hand, although rapidly frozen tissue, postfixed with cold Carnoy promises dimensional stability without introducing any postmortem and perfusion artifacts, it is good only for light microscopy. So far, it has not been usable for electron microscopy, since crystallization, especially during the early thawing phases, destroys much submicroscopic detail.

Figure 13.8 shows that one can determine the hematocrit in any specific blood vessel by counting "hits" in a 100-point grid.

Fig. 13.8. Blood in an interlobular artery of a dog's kidney. $\frac{1}{2}$ μm "section" with 100-point orthogonal lattice grid superimposed to obtain hematologic data. Width of grid = 33 μm.

The surface area of erythrocytes per unit volume of blood can be determined by intersection counts of traces of cell surfaces with grid lines. In the specific image shown in figure 13.8, the length of each individual line is $\ell = 33$ μm and the total length of all 20 lines together is $L = 660$ μm $\simeq \frac{2}{3}$ mm.

Also the number of cells per cubic millimeter of blood can be determined stereologically by formula 5.3, after $\bar{D}$ has been determined.

Knowledge of the surface area per unit volume of blood s_V is important, since it is a determinant of diffusion. On the other hand, rouleaux formation (fig. 13.9) presents another problem—i.e., do the contact surfaces between individual RBC in a rouleau contribute to or do they eliminate potential diffusion surfaces? Is rouleaux formation, perhaps a mechanism for the inhibition of diffusion? This is a question for physiologists to answer.

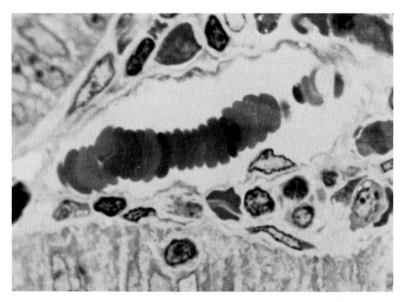

Fig. 13.9. A rouleaux in a venule of the colonic lamina propria. Should the mutual contact surfaces of the erythrocytes be included into the diffusion surface; or is rouleaux formation a mechanism for the reduction of diffusion?

Up to now quantitative blood analysis has been possible for peripheral blood drawn from a blood vessel only. Stereology permits such determinations in individual vessels during specific functional states.

For such determinations in larger animals, more complicated methods of sampling are required.

Glomerular Basement Membrane. The stereological analysis of the thickness of the glomerular basement membrane (GMB) has been discussed in Chapter 6 by *Elias*. It is not a simple matter to determine a range of its thickness, although it is very easy to decide, using the graphs presented by

Elias (1980) and in our Chapter 6 to find out whether the GMB is of constant thickness or not and what that thickness is at the thinnest place. A satisfactory method to determine range and distribution of thickness of a membrane that is almost structureless has not yet been found. Perhaps some nephrologist with a talent for mathematics will find this method soon.

This section, however, deals with different aspects of the GMB that could easily be quantified by conventional stereology. These are listed below.

1. Filtration surface area; this has been done in a coarse manner using human autopsy material from accidental deaths, in 3 μm paraffin sections, by *Elias and Hennig* (1967).
2. Is the GMB extensible?
3. Is the GMB elastic?

A sheet of parafilm is extensible (2), i.e., it can be stretched, but, when released will not return to its former less extensive area. Is the GMB of this nature? A rubber sheet is elastic (1) and, after being released from a stretched condition will snap back to its former size. Is the GMB elastic? In animal experiments, these questions could be answered very easily. Mice are small enough to produce practically instant freezing in liquid nitrogen or freon. Here is a possible method:

Use syngeneic mice, possibly litter mates. Lightly anesthetize each animal individually with nitrous oxide. The animal is put in a tall, narrow box with one wall of glass. In the lid are two holes. A rubber tube brings the gas into one hole. Air can escape through the other. The second hole can be closed with a cork when you believe the gas concentration to be high enough. When the mouse is asleep, which you can roughly observe through the glass window, it is removed from the anesthetic box and a corneal reflex test is tried. If necessary, anesthesia can be continued with an ether cone.

When the animal is completely asleep but still fully living, shave the abdomen, open the abdominal wall and clamp one renal vein with small hemostatic pincers. All blood vessels in the clamped kidney should be fully extended, but not beyond a natural condition, after 2 minutes.

Now evert both kidneys, letting them hang out of the abdominal incision and, with the abdomen wide open, immerse the entire animal in liquid freon cooled in liquid nitrogen. The mouse is rapidly frozen. Still in the liquid freon, use a chisel to remove both kidneys. Immerse them separately in supercooled Carnoy mixture as described previously.

After plastic embedding and differential stain for basement membrane, S_V can be determined in semithin "sections" ($\frac{1}{2}$ μm). S_V of the blood-engorged kidney can now be compared with S_V of the control (the contralateral kidney). If the former is larger than the latter, the GMB is extensible.

To decide whether it is also elastic, the above experiment should be modified: clamp or ligate the renal vein of the experimental kidney for two minutes. Then release the ligature and a few minutes after that, cut both the renal artery and the renal vein distal to the ligature. Do not cut the ureter. Squeeze the experimental kidney gently. Wait 2 minutes, then freeze and proceed as above. Compare S_V in all three conditions. Note that when determining S_V of the glomerular basement membrane by the formula $S_V = 2P/L$, L is *not* the combined length of all the test lines in the grid. It is only the combined length of test lines coextensive with the glomerulus. However, portions of grid lines which project into Bowman's space must be subtracted (fig. 13.10). There are more possible variations of this method. Much can be learned about the physical properties of the glomerular basement membrane.

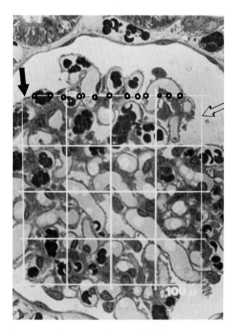

Fig. 13.10. Surface area determination of glomerular basement membrane by interaction counts P_L: Line segments falling outside of the glomerulus, as pointed out by arrows, must be subtracted from L.

Appendix I. Programs for Estimating Mean Caliper Diameter From Serial Sections*

The programs and their subroutines are organized according to their use. The first program MOPIN uses a Zeiss MOP AMO3 to enter either discrete parameter data or continuous X-Y data of traced section profiles. "MOPIN" creates a disk file which is assigned a name by the operator. The file name is evaluated by subroutines CKNAME and LEGAL, which check RT-11 filenames for proper format. Subroutines DATE and ASSIGN are intrinsic library functions to the operating system.

Subroutine INPTOP inputs the data file header information and comments at the top of the data file. "INPTOP" calls subroutines "COMMNT" which is used to write comment lines to the data file and "FIND" which is used to identify the type of parameters (discrete or continuous) passed to the file. Once the type of parameter is identified, then the appropriate subroutine DSCR for discrete parameters or "CONT" for continuous parameters is called. At the present time we have not written software for any other digitizer besides the Zeiss MOP AMO3.

The mean caliper diameter MCD program is explained conceptually in Chapter XI. After the program initializes parameters and maximum (MAX) and minimum (MIN) one-dimensional arrays, it reads a file name and calls subroutine "CKNAME" to evaluate it for proper format. The image (section) is then tested—to be sure it has not been rotated—by another program used for viewing the image. If the image has not been rotated, the file is read. Note that section spacing is also found within the file; it was entered in subroutine INPTOP of program MOPIN. Section spacing is in MOP units entered as μm per mm.

Next, subroutine GETSET is called, which inputs a block of X-Y data into X and Y arrays and puts the block identification number in ID.

The identification number (ID) for each block is used to update Z distance of each additional section. DO loops are used for X-axis and Y-axis rotations. The remainder of the MCD program is presented in detail in Chapter XI. A flow chart is also provided for the MCD program (A-I.1). Since the axes of rotation are orthogonal and each is rotated through 180°, for all possible combinations of the fixed rotations only the maximum x

---

* Programs written by Mr. E.F. Reus and Ms. N.K. Tyler

and minimum x for each rotation is needed to calculate the mean caliper diameter. The caliper diameter of each orientation is used to calculate the mean caliper diameter. The weighted factor (WD) is added because the coordinate transformations are biased for equal angle increments (the density of the rotations is not constant).

MEAN CALIPER DIAMETER PROGRAM

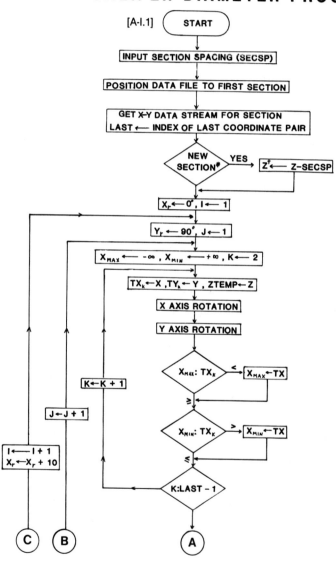

Programs for Estimating Mean Caliper Diameter From Serial Sections

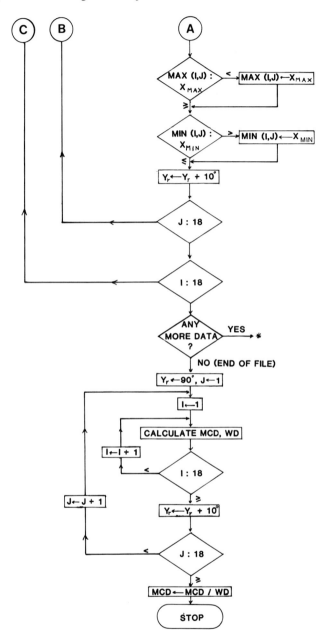

```
               PROGRAM MOPIN

C**********************************************************************
C    MOPIN.FOR                                         EFR 21-JUL-80   *
C                                                                      *
C      Program for reading ZEISS MOP AM03 data.                        *
C                                                                      *
C      File 4 is a disk file (default device) which will hold          *
C      the new data.                                                   *
C      File 3 is the serial port connected to the MOP.                 *
C                                                                      *
C      Subroutines used by MOPIN                                       *
C           CKNAME.OBJ                                                 *
C           FIND.OBJ                                                   *
C           INPTOP.OBJ                                                 *
C           DSCR.OBJ                                                   *
C           CONT.OBJ                                                   *
C                                                                      *
C**********************************************************************
           LOGICAL*1   DSKFILE(11),ERROR,ESC,FF,BUF(80)
           INTEGER     TYPE
           REAL        VERSION

C    constants:
           VERSION=2.001              ! program version.
           ESC=27                     ! ASCII escape character.
           FF=12                      ! ASCII form feed character.

C    initializations:
           TYPE=0                     ! type of data collected.

C    output title, open disk target file (4), open read port (3):
           WRITE(7,100) ESC,FF,VERSION
100        FORMAT(2A1,T12,'Zeiss MOP-3 data collection program
      *                  'Version ',F5.3///)
105        WRITE(7,110)
110        FORMAT('   Input name of new data file')
120        READ(5,130) BUF
130        FORMAT(80A1)
           CALL CKNAME(DSKFILE,BUF,80,ERROR)
           IF (ERROR) GOTO 120
           OPEN( UNIT=4, NAME=DSKFILE, TYPE='NEW', ERR=600 )
           CALL ASSIGN( 3, 'PC:', 3 )

C    read in file header information:
           CALL INPTOP(TYPE,DSKFILE)

C    input the file according to its type.
C         TYPE=1 : Discrete parameter data, one or more structures
C                  per block of data.
C         TYPE=2 : Continous X-Y data, one or more structures per
C                  block of data.

           IF( (TYPE.LT.1).OR.(TYPE.GT.2) ) GOTO 500       ! error?
           IF( TYPE.EQ.1 ) CALL DSCR(TYPE)
           IF( TYPE.EQ.2 ) CALL CONT(TYPE)

           ENDFILE 4
           CLOSE( UNIT=4 )

           STOP ' File saved... All done...'
```

```
C   TYPE out of bounds:
500      STOP ' Variable type out of bounds... From MOPIN...'

C   file open error:
600      PAUSE ' File open error... From MOPIN...'
         GOTO 105

         END
```

```
      SUBROUTINE CKNAME(FILE,BUF,N,ERROR)
C***********************************************************************
C   Routine CKNAME                                    EFR 16-may-80    *
C                                                                      *
C      This subroutine ckecks RT-11 filenames for proper format.        *
C   The RT-11 filenames considered do not have the device kind          *
C   listed, only file name and file type.  The following format         *
C   is allowed:                                                         *
C                                                                      *
C         name.typ                                                      *
C                                                                      *
C      where:     name    : one to six characters. Valid chars are      *
C                           upper case letters (A-Z) or digits          *
C                           (0-9).                                      *
C                   .     : name/type seperator. There must be a        *
C                           period between the filename and file-       *
C                           type.                                       *
C                  typ    : zero to three characters defining the       *
C                           filetype. Valid chars are upper case        *
C                           letters or digits.                          *
C                                                                      *
C      Soubroutine LEGAL is called to check the validity of each        *
C   character in the filename. If a filename is found invalid           *
C   CKNAME returns with ERROR set to true.                              *
C                                                                      *
C***********************************************************************
        LOGICAL*1    FILE(11),ERROR,BUF(N),ILLEGAL,PERIOD
        INTEGER      ISTART,IEND,NP,N

        NP=0              ! number of periods encountered.
        FILE(11)=0        ! file name ends with ASCII null.
        PERIOD=.FALSE.    ! initialise period flag.
        ERROR=.FALSE.     ! syntax error flag.

C   initialize FILE to blanks:
        DO 50 I=1,10
50         FILE(I)=' '

C   determine start and end points of filename in BUF:
        DO 100 I=1,N
           IF( BUF(I).NE.' ' ) GOTO 110        ! check for character.
100     CONTINUE
        ERROR=.TRUE.                           ! empty buffer
        RETURN

110     ISTART=I
        DO 120 I=1,N
           J=N+1-I
           IF( BUF(J).NE.' ' ) GOTO 130        ! check for character.
120     CONTINUE

130     IEND=J
        IF( (IEND-ISTART).GT.9 ) GOTO 200      ! check filename length.

C   transfer filename from BUF to FILE:
        J=0
        DO 180 I=ISTART,IEND
           CALL LEGAL(BUF(I),ILLEGAL,PERIOD,NP)
           IF( ILLEGAL ) GOTO 200              ! illegal char found.
           IF( PERIOD.AND.(I.EQ.ISTART) ) GOTO 200
```

```
              IF( PERIOD.AND.(I.GT.(ISTART+6)) ) GOTO 200
              IF( PERIOD.AND.((IEND-I).GT.3) ) GOTO 200
              J=J+1
              FILE(J)=BUF(I)
180       CONTINUE
          RETURN               ! normal termination.

C  syntax error encountered:
200       WRITE(7,220)
220       FORMAT('   Syntax error')
          ERROR=.TRUE.
          RETURN               ! return with syntax error.
          END


          SUBROUTINE LEGAL(CHAR,ILLEGAL,PERIOD,NP)


C*********************************************************************
C  LEGAL                                                             *
C                                                                    *
C    Called by CKNAME to chaek for valid filename characters.        *
C                                                                    *
C    Checks to see that CHAR is a legal character. A legal           *
C  character is: upper case letter (A-Z); digit (0-9);               *
C  or a period if NP is not greater than one. If an                  *
C  illegal character is found ILLEGAL is set, otherwise              *
C  it stays false.                                                   *
C                                                                    *
C*********************************************************************
          LOGICAL*1   CHAR,ILLEGAL,PERIOD
          INTEGER     NP

          ILLEGAL=.FALSE.
          PERIOD=.FALSE.

C  check if character is a period:
          IF( CHAR.NE.'.') GOTO 100
          PERIOD=.TRUE.
          NP=NP+1
          IF( NP.GT.1 ) ILLEGAL=.TRUE.    ! name can have only one period.
          RETURN                          ! exit for period.

C  check if character is upper case letter (A-Z):
100       DO 120 I=65,90
              IF( CHAR.EQ.I ) RETURN      ! exit for legal letter.
120       CONTINUE

C  check if character is digit (0-9):
          DO 140 I=48,57
              IF( CHAR.EQ.I ) RETURN      ! exit for legal digit.
140       CONTINUE

          ILLEGAL=.TRUE.
          RETURN                          ! exit for illegal character.

          END
```

```
      SUBROUTINE INPTOP(TYPE,DSKFILE)
C***********************************************************************
C    INPTOP.FOR                                    EFR 21-JUL-80    *
C                                                                   *
C    Called by program MOPIN, this subroutine inputs the data       *
C    file header information and comments at the top of the data    *
C    file.                                                          *
C    Parameters (passed by calling routine):                        *
C        DSKFILE : Disk file name store in 11-character array       *
C                  ( 11-element logical*1 array).                   *
C        TYPE    : Type if data to be sent by the digitizer.        *
C                  Type can have four values:                       *
C                     1 : Discrete parameters sent from MOP, one    *
C                         structure per block of data.              *
C                     2 : Discrete parameters sent from MOP, one    *
C                         or more structures per block of data.     *
C                     3 : Continous streams of X-Y data, only one   *
C                         structure per block of data.              *
C                     4 : Continous streams of X-Y data, one or     *
C                         more structures per block of data.        *
C                                                                   *
C    Subroutines called by INPTOP:                                  *
C        COMMNT, FIND                                               *
C                                                                   *
C***********************************************************************
      LOGICAL*1   BUF1(60),LDATE(9),DSKFILE(11),OPTION
      INTEGER     TYPE
      REAL        SECSP

C   data file title:
      WRITE(7,140)
140   FORMAT(3X,'Input data file title')
      READ(5,150) BUF1
150   FORMAT(60A1)
      WRITE(4,160) BUF1
160   FORMAT('T  Title',T17,':',60A1)

C   investigator:
      WRITE(7,170)
170   FORMAT(3X,'Input investigator name')
      READ(5,150) BUF1
      WRITE(4,180) BUF1
180   FORMAT('T  Investigator :',60A1)

C   today's date:
      CALL DATE(LDATE(1))
      WRITE(4,190) LDATE
190   FORMAT('T  Date',T17,':',9A1)

C   disk number:
      WRITE(7,200)
200   FORMAT(3X,'Input disk number')
      READ(5,150) BUF1
      WRITE(4,210) BUF1
210   FORMAT('T  Disk no.      :',60A1)

C   data file name:
      WRITE(4,220) (DSKFILE(I),I=1,10)
220   FORMAT('T  File name     :',10A1)
```

```
C   data strurture questions (determine data TYPE):
        WRITE(7,280)
280     FORMAT(3X,'Will there be contigous X-Y data? (Y/N)')
        READ(5,290) BUF1
290     FORMAT(60A1)
        CALL FIND(OPTION,BUF1,60)
        TYPE=1
        IF( OPTION.EQ.'Y' ) TYPE=2

500     WRITE(4,520) TYPE
520     FORMAT('T   Data type',T17,':MOP',I1)

C   serial section spacing:
530     WRITE(7,540)
540     FORMAT('   Input serial section spacing (MOP units)')
        READ(5,560,ERR=530) SECSP
560     FORMAT(F7.0)
        WRITE(4,580) SECSP
580     FORMAT('T   Sec spacing',T17,':',F7.1)

C   input any comments
        CALL COMMNT

        RETURN   ! done...

        END
```

```
          SUBROUTINE FIND(MOD,BUF,N)
C*******************************************************************
C                                                                   *
C     This routine searches N-character BUF for the first           *
C  non-blank character and puts it in MOD.                          *
C     If no non-blank characters are found in BUF then MOD          *
C  returns with "blank" value.                                      *
C                                                                   *
C*******************************************************************
          LOGICAL*1 MOD,BUF(N)

          MOD=' '
          DO 200 I=1,N
             IF( BUF(I).EQ.' ' ) GOTO 200
             MOD=BUF(I)
             RETURN
200       CONTINUE

          RETURN
          END
```

```
        SUBROUTINE COMMNT
C*****************************************************************
C   COMMNT.FOR                                 EFR 18-JUL-80   *
C                                                              *
C    Write comment lines to file 4. Return if two contigous    *
C  slashes are encountered.                                    *
C                                                              *
C*****************************************************************

        LOGICAL*1  BUF(75)

        WRITE(7,100)
100     FORMAT('    Input any comments (type "//" to end comment mode)')

120     READ(5,140) BUF
140     FORMAT(75A1)

        DO 160 I=1,74
           IF( (BUF(I).EQ.'/').AND.(BUF(I+1).EQ.'/') ) GOTO 200
160     CONTINUE

        WRITE(4,180) BUF
180     FORMAT('C  ',75A1)

        GOTO 120              ! read more comments...
C end of comment mode:
200     IF( I.LE.1 ) RETURN   ! // type at start of line, no
C                             ! comments left.

        WRITE(4,180) (BUF(J),J=1,I-1)   ! output last comment line.

        RETURN
        END
```

```
          SUBROUTINE DSCR(TYPE)

C*******************************************************************
C   DSCR.FOR                                         EFR 21-JUL-80   *
C                                                                    *
C     Called by MOPIN, this subroutine reads discrete                *
C   measurements (standard parameters measured by the ZEISS          *
C   MOP) to a disk data file (file 4).                               *
C                                                                    *
C*******************************************************************

          LOGICAL*1   BUF1(60),DATA(130),BELL,ESC,FF,ERROR
          INTEGER     BREAKPOINT,TYPE,ID

          BELL=7              ! terminal bell tone (ASCII).
          FF=12               ! form feed character.
          ESC=27              ! escape character.
          BREAKPOINT=66       ! end of data line marker.

C   block ID:
300       WRITE(7,310) ESC,FF,BELL
310       FORMAT(3X,3A1,40(' ')/
     *           $'Input block ID (type // to end input):')
312       READ(5,420) BUF1
          IF( (BUF1(1).EQ.'/').AND.(BUF1(2).EQ.'/') ) RETURN
          CALL CONVRT(BUF1,10,ID,ERROR)
          IF (ERROR) GOTO 312
314       WRITE(4,315) ID
315       FORMAT('H Section ID',T17,':',I5)

C   input data:
320       WRITE(7,325)
325       FORMAT(3X,'Input a data structure')
          READ(3,330) DATA
330       FORMAT(130A1)

C   will data fit on one line?
          DO 340 J=1,130
             K=131-J
             IF( DATA(K).NE.' ' ) GOTO 350
340       CONTINUE
350       IF( K.GT.BREAKPOINT ) GOTO 370

C   single line of data:
          WRITE(4,360) (DATA(I),I=1,K)
360       FORMAT('D ',77A1)
          GOTO 390

C   two lines of data:
370       WRITE(4,360) (DATA(I),I=1,BREAKPOINT)
          WRITE(4,380) (DATA(I),I=BREAKPOINT+1,K)
380       FORMAT('D ',75A1)

C   more data for this block? more blocks?
390       WRITE(7,400) BELL
400       FORMAT($3X,A1,'Any more data for this section? (Y/N):')
          READ(5,420) BUF1
420       FORMAT(60A1)
          CALL FIND(OPTION,BUF1,60)
          IF( (OPTION.NE.'Y').AND.(OPTION.NE.'N') ) GOTO 390
          IF( OPTION.EQ.'Y' ) GOTO 314
          GOTO 300

          END
```

```
            SUBROUTINE CONT(TYPE)
C*****************************************************************
C   CONT.FOR                                    EFR 21-JUL-80    *
C                                                                *
C   This subroutine, called by MOPIN, reads continous streams    *
C   of X-Y coordinate data from the MOP. The MOP must be         *
C   programed so that it two kinds of data for this routine      *
C   to work: X-Y data, and AREA (or any other single valued      *
C   parameter measured by the MOP) must be sent. This routine    *
C   will hang in a infinite read loop if only X-Y coordinates    *
C   are sent.                                                    *
C      Example MOP program:                                      *
C                  <SET>           ! start program.              *
C                  <SEND>          ! send data to interface.     *
C                  <X;Y>           ! set X-Y coordinates.        *
C                  <AREA>          ! set area measurement.       *
C                  <ENTER>         ! end program.                *
C                                                                *
C   Note: CONT doesn't work if the baud rate of the serial       *
C   port is greater than 1200 baud.                              *
C                                                                *
C*****************************************************************
            LOGICAL*1   BUF1(60),HOLD(500,34),OPTION,BELL,
        *               ESC,FF,ERROR
            INTEGER     TYPE,N,JSTART,ID

C   constants:
            BELL=7              ! terminal bell tone (ASCII).
            FF=12               ! form feed character.
            ESC=27              ! escape character.

C   start section count (number of blocks):
100         N=0

C   block ID:
110         WRITE(7,120) ESC,FF,BELL
120         FORMAT(1X,3A1,40(' ')/$
        *         1X,' Input block ID (Type // to end input):')
140         READ(5,150)   BUF1
150         FORMAT(60A1)
            IF( (BUF1(1).EQ.'/').AND.(BUF1(2).EQ.'/') ) RETURN
            CALL CONVRT(BUF1,10,ID,ERROR)
            IF (ERROR) GOTO 140
170         WRITE(4,200) ID
200         FORMAT('H  Section ID',T17,':',I5)

C   input data:
510         WRITE(7,512)
512         FORMAT('   Input a data structure...')
515         N=N+1
            READ(3,520) (HOLD(N,J),J=1,31)
520         FORMAT(31A1)
            IF( HOLD(N,19).EQ.' ' ) GOTO 550
            GOTO 515

C   done reading block, put data to disk:
550         DO 560 I=1,N

C           find start of data in Jth line (find first '.'):
            DO 552 JSTART=1,34
```

```
                   IF( HOLD(I,JSTART).EQ.'.' ) GOTO 554
552        CONTINUE

           GOTO 560        ! skip blank line.

C          write lint to disk file:
554        WRITE(4,555) (HOLD(I,J),J=JSTART,31)
555        FORMAT('D  ',31A1)
560        CONTINUE


C  more data for this block?
620        WRITE(7,640) BELL
640        FORMAT($1X,A1,' Any more images on this section? (Y/N):')
           READ(5,150) BUF1
               CALL FIND(OPTION,BUF1,60)
               IF( OPTION.EQ.'N' ) GOTO 100
               N=0
               GOTO 170

           END
```

Programs for Estimating Mean Caliper Diameter From Serial Sections

```
            PROGRAM MCD
C
C****************************************************************
C     MCD.FOR                                   efr 7-Aug-80   *
C                                               REV. NKT DEC-80 *
C                                                                *
C     This program is used to calculate the mean caliper         *
C     diameter of a three dimensional object. The data file      *
C     containing the object must be of type "MOP2" and not have  *
C     been rotated by the VIEW program.                          *
C     The program calculates the mean caliper by going through   *
C     a set of predetermined rotations. NXR is the number of     *
C     X-axis rotations and DX is the angular increment in radians.*
C     For the Y-axis, NYR is the number of rotations preformed and*
C     DY is its angular increment.                               *
C                                                                *
C****************************************************************
            LOGICAL*1  DATFLE(11),DONE,BUF(80),TEST
            INTEGER    N,ID,OLDID,NXR,NYR,LAST
            REAL       X(500),Y(500),TX(500),TY(500),XR,YR,ZR,
     *                 ZPLANE,MCD,WD,DY,DX,SECSP,VERSION,
     *                 SINYR,TYPE,MAX(18,18),MIN(18,18),XMAX,XMIN,
     *                 SINXR,COSXR,COSYR,ZTEMP,YTEMP

            XR=0.0            ! initialize x-axis rotation angle.
            YR=0.0            ! initialise z-axis rotation angle.
            ZR=0.0            ! z-axis rotation won't be used.
            WD=0.0            ! weighted devisor for finding the av. MCD.
            OLDID=-1          ! ID number of last block input...
            DY=10.0/57.29577  ! y-axis incremental rotation (10 deg).
            DX=DY             ! x-axis inc. rot.
            ZPLANE=0.0        ! current Z-coordinate (depth into sec).
            VERSION=2.001     ! program version number.
            N=500             ! length of X/Y arrays.
            NXR=18            ! number of x rotations.
            NYR=18            ! number of y rotations.
            MCD=0.0           ! mean caliper diameter.
            WD=0.0            ! weighted number of diameters measured.

C     initialize the minimum/maximum arrays (for rotation bounds):
            DO 50 I=1,NXR
              DO 40 J=1,NYR
                MAX(I,J)=-10000.
                MIN(I,J)=10000.0
40            CONTINUE
50          CONTINUE

C     output prog title, open data file:
            WRITE(7,100) VERSION
100         FORMAT(1H1,T12,'Mean Caliper Diameter Program
     *                    'Version ',F5.3///)
            WRITE(7,110)
110         FORMAT($'   Input data file name:')
120         READ(5,130) BUF
130         FORMAT(80A1)
            CALL CKNAME(DATFLE,BUF,80,ERROR)
            IF (ERROR) GOTO 120
            OPEN( UNIT=3, NAME=DATFLE, TYPE='OLD', ERR=140 )
            GOTO 145
140         STOP ' Hardware file open error called... '
```

```
C   calculate MCD by predetermined angles:
C       check that image has not been rotated:
145     DO 150 I=1,4
            READ(3,180) BUF(1)
150     CONTINUE
        READ(3,155) XR,YR,ZR
155     FORMAT(51X,3(7X,F7.0))
        TEST= ((XR.EQ.0.0).AND.(YR.EQ.0.0)).AND.(ZR.EQ.0.0)
        IF (TEST) GOTO 170
            WRITE(7,160)
160         FORMAT(///'    I caught you!! This image file has already'/
     *                 '    been rotated. You can only use image files that'/
     *                 '    have been created by MOPIN (not ones created by'/
     *                 '    VIEW). So use the original data file you made'/
     *                 '    up when you digityzed the image.'/)
            STOP ' Bye...'

C       set section spacing and position file to start of data:
170     READ(3,174) TYPE
174     FORMAT(17X,A4)
        IF(TYPE.NE.'MOP2') STOP'Data file of wrong type...'
        READ(3,176) SECSP
176     FORMAT(17X,F7.0)
178     READ(3,180) BUF(1)
180     FORMAT(A1)
        IF( BUF(1).NE.'H' ) GOTO 178
        BACKSPACE 3


C       input the next section:
200     CALL GETSEC(X,Y,N,ID,DONE)

        WRITE(7,220) ID
220     FORMAT(/'    Starting section #',I5)

        LAST=IFIX(X(1))+1               ! index of last coord in X/Y.

C       is this a new section deeper into the tissue or is
C       the new x/y data on the same section as the last?
        IF( (OLDID.NE.ID).AND.(OLDID.NE.-1) ) ZPLANE=ZPLANE-SECSP
        OLDID=ID

C       in this do loop, I iterates for the X-axis rotations
C       and J iterates for the Y-axis rotations. Note the
C       weighting factor: sin(YR) for the second DO loop set.
        XR=0.0
        DO 350 I=1,NXR

            SINXR=SIN(XR)               ! pre-calculate X sines/cosines.
            COSXR=COS(XR)
            YR=-1.570795                ! start Y rotation at -90 degrees...
            DO 300 J=1,NYR

                XMAX=-1000.0            ! initialize x bounds.
                XMIN=10000.0

                SINYR=SIN(YR)
                COSYR=COS(YR)
                DO 280 K=2,LAST-1

                    TX(K)=X(K)          ! don't want to modify original data.
                    TY(K)=Y(K)
                    ZTEMP=ZPLANE
```

```
                    IF( XR.EQ.0.0 ) GOTO 270
                    YTEMP=TY(K)
                    TY(K)=TY(K)*COSXR+ZTEMP*SINXR
                    ZTEMP=ZTEMP*COSXR+YTEMP*SINXR

270                 IF( YR.EQ.0.0 ) GOTO 275
                    TX(K)=TX(K)*COSYR-ZTEMP*SINYR

275                 IF( XMAX.LT.TX(K) ) XMAX=TX(K)
                    IF( XMIN.GT.TX(K) ) XMIN=TX(K)

280            CONTINUE

                    IF( MAX(I,J).LT.XMAX ) MAX(I,J)=XMAX
                    IF( MIN(I,J).GT.XMIN ) MIN(I,J)=XMIN
                    YR=YR+DY              ! next Y-rotation (rads)...
300            CONTINUE
               XR=XR+DX                    ! next X-rotation (rads)...
350       CONTINUE

          IF(.NOT.DONE) GOTO 200

          YR=-1.570795
          DO 450 J=1,NYR
               SINYR=SIN(YR)
               DO 400 I=1,NXR
                    MCD=MCD+ABS(MAX(I,J)-MIN(I,J))*ABS(SINYR)
                    WD=WD+ABS(SINYR)
400            CONTINUE
               YR=YR+DY
450       CONTINUE

          MCD=MCD/WD

C    output results to console:
          WRITE(7,500) MCD
500       FORMAT(///' Mean caliper diameter =',F12.3/)
          CLOSE( UNIT=3 )

C    save intermediate calculations (for testing only):
C         OPEN( UNIT=3, NAME='MCDTST.TST', TYPE='NEW' )
C         WRITE(3,520) (DATFLE(J),J=1,10),MCD
C520      FORMAT(////'      Data file : ',10A1//
C    *                    '      Mean Caliper diameter =',F12.3//
C    *                    ' X-rot    Y-rot   (in degrees)')
C         DO 600 I=1,NXR
C              DO 550 J=1,NYR
C                   WRITE(3,530) I-1,J-1,MAX(I,J)-MIN(I,J)
C530               FORMAT(/3X,I2,'O',5X,I2,'O',5X,'X- max dia =',
C    *                    F10.3,2(3X,F10.1))
C550           CONTINUE
C600      CONTINUE
C         CLOSE (UNIT=3)

          STOP 'All Done...'
          END
```

```
      SUBROUTINE CKNAME(FILE,BUF,N,ERROR)
C*********************************************************************
C   Routine CKNAME                                    EFR 16-may-80   *
C                                                                     *
C     This subroutine ckecks RT-11 filenames for proper format.       *
C   The RT-11 filenames considered do not have the device kind        *
C   listed, only file name and file type.  The following format       *
C   is allowed:                                                       *
C                                                                     *
C       name.typ                                                      *
C                                                                     *
C   where:       name    : one to six characters. Valid chars are     *
C                          upper case letters (A-Z) or digits         *
C                          (0-9).                                     *
C                 .      : name/type seperator. There must be a       *
C                          period between the filename and file-      *
C                          type.                                      *
C                typ     : zero to three characters defining the      *
C                          filetype. Valid chars are upper case       *
C                          letters or digits.                         *
C                                                                     *
C     Soubroutine LEGAL is called to check the validity of each       *
C   character in the filename. If a filename is found invalid         *
C   CKNAME returns with ERROR set to true.                            *
C                                                                     *
C*********************************************************************
        LOGICAL*1   FILE(11),ERROR,BUF(N),ILLEGAL,PERIOD
        INTEGER     ISTART,IEND,NP,N

        NP=0              ! number of periods encountered.
        FILE(11)=0        ! file name ends with ASCII null.
        PERIOD=.FALSE.    ! initialise period flag.
        ERROR=.FALSE.     ! syntax error flag.

C   initialize FILE to blanks:
        DO 50 I=1,10
50          FILE(I)=' '

C   determine start and end points of filename in BUF:
        DO 100 I=1,N
            IF( BUF(I).NE.' ' ) GOTO 110        ! check for character.
100     CONTINUE
        ERROR=.TRUE.                            ! empty buffer
        RETURN

110     ISTART=I
        DO 120 I=1,N
            J=N+1-I
            IF( BUF(J).NE.' ' ) GOTO 130        ! check for character.
120     CONTINUE

130     IEND=J
        IF( (IEND-ISTART).GT.9 ) GOTO 200       ! check filename length.

C   transfer filename from BUF to FILE:
        J=0
        DO 180 I=ISTART,IEND
            CALL LEGAL(BUF(I),ILLEGAL,PERIOD,NP)
            IF( ILLEGAL ) GOTO 200              ! illegal char found.
            IF( PERIOD.AND.(I.EQ.ISTART) ) GOTO 200
```

Programs for Estimating Mean Caliper Diameter From Serial Sections 219

```
              IF( PERIOD.AND.(I.GT.(ISTART+6)) ) GOTO 200
              IF( PERIOD.AND.((IEND-I).GT.3) ) GOTO 200
              J=J+1
              FILE(J)=BUF(I)
180       CONTINUE
          RETURN             ! normal termination.

C   syntax error encountered:
200       WRITE(7,220)
220       FORMAT('   Syntax error')
          ERROR=.TRUE.
          RETURN             ! return with syntax error.
          END



          SUBROUTINE LEGAL(CHAR,ILLEGAL,PERIOD,NP)


C*******************************************************************
C   LEGAL                                                          *
C                                                                  *
C     Called by CKNAME to check for valid filename characters.     *
C                                                                  *
C     Checks to see that CHAR is a legal character. A legal        *
C   character is: upper case letter (A-Z); digit (0-9);            *
C   or a period if NP is not greater than one. If an               *
C   illegal character is found ILLEGAL is set, otherwise           *
C   it stays false.                                                *
C                                                                  *
C*******************************************************************
          LOGICAL*1    CHAR,ILLEGAL,PERIOD
          INTEGER      NP

          ILLEGAL=.FALSE.
          PERIOD=.FALSE.

C   check if character is a period:
          IF( CHAR.NE.'.') GOTO 100
          PERIOD=.TRUE.
          NP=NP+1
          IF( NP.GT.1 ) ILLEGAL=.TRUE.   ! name can have only one period.
          RETURN                         ! exit for period.

C   check if character is upper case letter (A-Z):
100       DO 120 I=65,90
              IF( CHAR.EQ.I ) RETURN     ! exit for legal letter.
120       CONTINUE

C   check if character is digit (0-9):
          DO 140 I=48,57
              IF( CHAR.EQ.I ) RETURN     ! exit for legal digit.
140       CONTINUE

          ILLEGAL=.TRUE.
          RETURN                         ! exit for illegal character.

          END
```

A Guide to Practical Stereology

```fortran
              SUBROUTINE GETSEC(X,Y,N,ID,DONE)

C*******************************************************************
C    GETSEC.FOR                                    EFR 22-JUL-80   *
C                                                                  *
C    Input a block of X-Y data to X and Y arrays, and put the      *
C    block identification number in ID. This subroutine assumes    *
C    that the data file (file 3) is positioned to the start of     *
C    the next block to be read. When GETSEC sets to eof, DONE is   *
C    set true and GETSEC returns with the last set (last block)    *
C    of data.                                                      *
C       Parameters (passed by calling routine):                    *
C                                                                  *
C         N       : Length of X-Y arrays.                          *
C                                                                  *
C       Parameters (returned by GETSEC):                           *
C                                                                  *
C         X       : Real array of N length holds list of X         *
C                   coordinates.                                   *
C         Y       : Real array of N length holds a list of the Y   *
C                   coordinates.                                   *
C         ID      : Integer identification number of the block     *
C                   being read.                                    *
C         DONE    : Logical*1 end of file flag (set to true on     *
C                   eof).                                          *
C                                                                  *
C*******************************************************************
          LOGICAL*1  DONE
          INTEGER    N,ID,NEXTID,ICOUNT
          REAL       CX,CY,X(N),Y(N)

          DONE=.FALSE.              ! initialize eof flag.
          ICOUNT=1                  ! start count of coordinates input.

C    initialize X/Y arrays:
          DO 50 J=1,N
              X(J)=0.0
              Y(J)=0.0
50        CONTINUE

C    input block header:
          READ(3,100) ID
100       FORMAT(17X,I5)

C    input the X-Y data:
120       ICOUNT=ICOUNT+1
          READ(3,140,ERR=200) X(ICOUNT),CX,Y(ICOUNT),CY
140       FORMAT(2X,2(F12.0,1X,F2.0))

C    output to console (for testing only):
C         WRITE(7,160) ICOUNT,X(ICOUNT),CX,ICOUNT,Y(ICOUNT),CY
C160      FORMAT(' X(',I2,')=',F7.3,'    CX=',F4.0,'    Y(',I2,')=',
C    *          F7.3,'    CY=',F4.0)

C    check Y-code, if Y-code equals zero then end of block has
C    been reached:
          IF( CY.NE.0.0 ) GOTO 120

C    we have come to the end of this block. Set the first element
C    in X and Y arrays to the number of coordinates they contain,
C    and check for eof:
```

```
              X(1)=FLOAT(ICOUNT)-2.0
              Y(1)=FLOAT(ICOUNT)-2.0
C     output number of data points (for testing only):
C             WRITE(7,170) X(1),Y(1),ID
C170          FORMAT(//'    Lengths:  X(1)=',F5.1,7X,'Y(1)=',F5.1/
C         *             '    Done with block ',I5)
C             PAUSE ' GETSEC...'

              READ(3,175,END=180) NEXTID       ! have we come to eof?
175           FORMAT(17X,I5)

              BACKSPACE 3
              RETURN          ! normal exit, not eof.

180           DONE=.TRUE.
              RETURN          ! normal exit, eof.
C     syntax error on reading X-Y data:
200           BACKSPACE 3
              READ(3,220) BUF
220           FORMAT(40A1)
              WRITE(7,240) ID,BUF
240           FORMAT('    Syntax error on reading X-Y coordinates.'/
          *          '    Error occured in block #',I5//
          *          '    Field in question: [',40A1,']')
              STOP ' Error encountered in subroutine GETSEC...'

              END
```

Appendix II. Programs for Estimating Stereological Values of Mammalian Lung and Heart

*The Mammalian Lung**

The stereology data entry program IN1 is used for on-line entry and tallying of point and intersection counts from either the light microscope (LM) or electron microscope (EM). It creates a disk file and requests the type of data to be entered: LM0 corresponds to level 1 of figure 12.1, LM2 to level 2, and EM1 to level 4. According to the data type, keys are designated for the appropriate structural components as listed in the program. After file title, investigator and disk number are entered, the program calls subroutine GRDTYP, which provides information about 6 possible test grids. The six test grids are described in detail in Appendix I of *Weibel* (1979). GRDTYP calls subroutines GETMOD, WRIFOR, FIND and CHECK and provides information about test point, lines, and areas of each grid type.

Next comment lines are called for. Subsequently, an animal identification number, comments concerning the animal, a magnification factor and lung volume are entered. Point counts and intersections can then be entered according to field, block and animal order. The stereological data entry program (IN2) is used for entering point and intersection counts that have already been tallied. It is very similar to the on-line program and it has the capability of adding data to old files pursuant to achieving the statistical requirements of an adequate sample. When entering new data, subroutine DEFTYP is called to define the label set and number of tally variables for each type of data (LM0, LM2 or EM1). Then subroutine FLINFO enables the investigator to enter all information concerning the file. FLINFO calls subroutine GRDTYP. Subroutine RTSD is called, which allows entry of tallied block data. RTSD calls subroutine "CONVERT" which converts an alpha string of digits to an integer.

---

* Programs written by Mr. E.F. Reus and Ms. N.K. Tyler

If new data are added to an old data file, subroutines MAKLST (makes a list of animals in old file), MOVTOP (transfers file header information to new file and defines labels), CONVRT (converts an alpha string of digits to an integer), MOVAN (moves all blocks of data from an old file to newfile), and RDANML (allows input of tallied data for an animal) are called. Both aspects (new data and new data added to old data files) continue until all data are entered.

Calculation of Stereological Data Program CALC

Light microscopic and transmission electron microscopic tallies from the IN1 and IN2 programs are used to calculate volumetric and surface densities along with specific thicknesses of pulmonary compartments. A-II.1 is a flow chart of the calculation program, which summarizes its general organization and operation.

The subroutines called by the calculation program (CALC) include: (a) CKNAME, which checks RT-11 filenames for proper format; (b) TOPFL, which reads the title and creates an output file header; (c) RBLOCK, which reads a block of tally data for processing; (d) CBLOCK, which performs data calculations for block totals; (e) ADDGRP, which adds a block of data for an animal to group results; (f) GRPRES, which outputs group results (means and standard deviations) to an output file; and (g) LMO, LM2, and EM1, which perform data calculations for levels 1, 2 and 4 of stereological data respectively for each animal (fig. 12.1).

*The Mammalian Heart**

Two programs are described which estimate (a) capillary profile densities after correction for orientation and a size distribution of capillaries normal to the plane of sectioning (with axial ratios of 1/1) and (b) stereological parameters from point and intersection counts at levels 1–4 (fig. 12.8). Although not included herein, a program that computes the group means, standard deviations, and outputs the results of stereological data can be written from a modification of the *Calculation of Stereological Data Program,* which is used for lung tissue (A-II.1).

---

* Programs written by M.J. Inderbitzen

The Capillary Data Program was developed to reduce raw capillary data to a more manageable form. The program has three primary functions: first, the entering of capillary axial ratios; second, the generation of a size distribution of capillaries with a 1/1 axial ratio; third, the determination of capillary profile densities.

The intersection of capillaries results in various shapes of ellipses, depending on orientation. The axial ratio is the ratio of the largest to the smallest diameter. Capillaries that have been intersected perpendicular to their longitudinal axis have a 1/1 axial ratio. These 1/1 ratios are grouped according to size, giving a distribution based on diameter. The capillary profile density for a given field is the mean axial ratio for that field multiplied by the number of capillaries and divided by the test area.

The individual ratios are not needed for further processing, so only the number of capillaries, sums of ratios, the frequency distribution, and the capillary profile density are stored in the output file along with other file characteristics. Since mistakes may be made while entering the individual capillary axial ratios, the program checks the total number entered with the number specified at the beginning of the field. Should a mistake occur, the field is reentered. A summary of this program is provided in a flow chart (A-II.2).

The *Stereological Data Entry Program* was developed to input and store stereological data, which will be used to calculate various stereological parameters and associated statistics. Volumetric and surface densities of components of the heart are the primary quanties to be determined.

Within each tissue sample, there are a number of blocks, and within each block, a number of fields. A variable number of fields and blocks is accepted by the program, including cases where a block or field has been skipped entirely. Sampling may be done only on a light microscopic level, or, for more definitive estimates, on both light and electron microscopic levels.

Different grids may be used to collect the data, but only one grid type is allowed for all levels of a particular tissue sample. The grids determine length and area variables. The data entry process can be stopped in the middle of a set of data and then continued at a later time. This makes data entry easier, since there is a large amount of data for each tissue sample. A summary of this program is provided in a flow chart (A-II.3).

A Guide to Practical Stereology

CALCULATION OF STEREOLOGICAL DATA PROGRAM

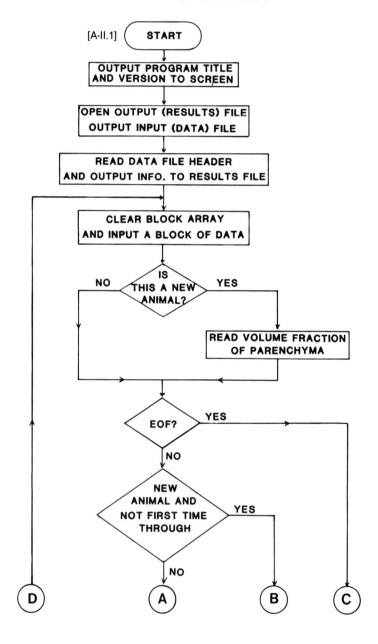

Programs for Estimating Stereological Values of Mammalian Lung and Heart

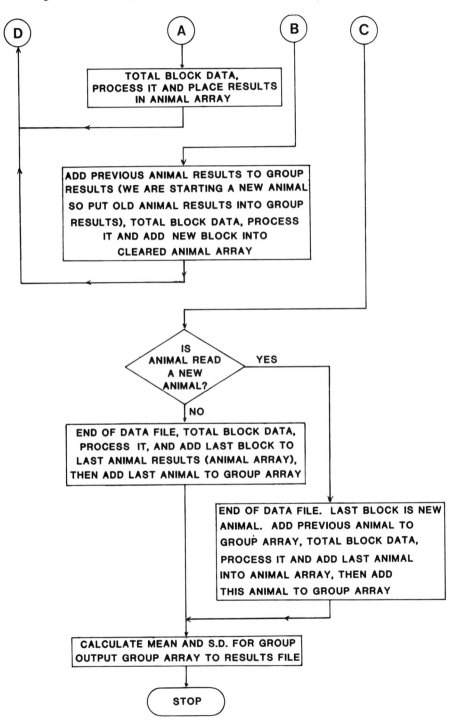

A Guide to Practical Stereology

CAPILLARY DATA PROGRAM

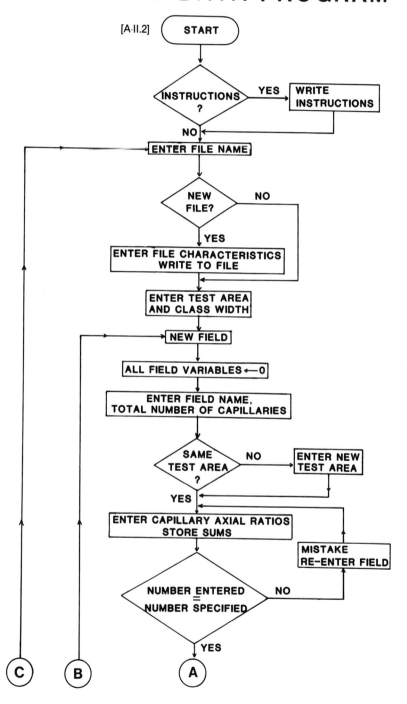

Programs for Estimating Stereological Values of Mammalian Lung and Heart

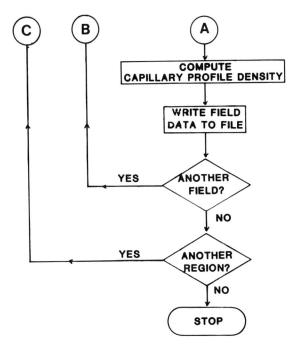

STEREOLOGICAL DATA ENTRY PROGRAM

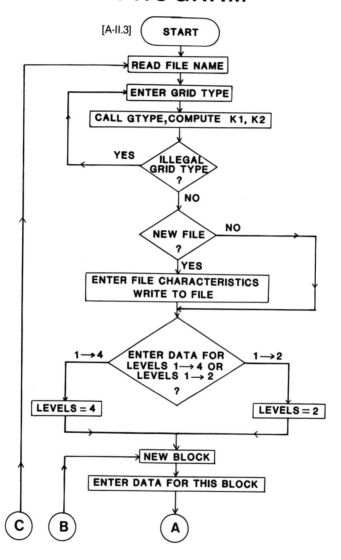

Programs for Estimating Stereological Values of Mammalian Lung and Heart

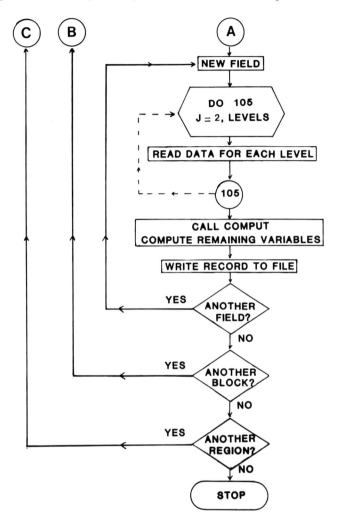

Stereology Data Entry Program IN1

```
              PROGRAM IN1
C***********************************************************************
C   Stereology Data Entry Program.                     EFR 6/Feb/80    *
C                                                      NKT 2-JUN-81    *
C                                                                      *
C     This program is used for entering and tallying stereology data   *
C   collected from either Light Microscope (LM) or Electron Microscope *
C   (EM).                                                              *
C***********************************************************************

      C   variables:
              LOGICAL*1    DSKFILE(11),CHAR(26),LDATE(9),BUF1(60),BUF2(76),
           *              OPTION,DATA(80),DONE,BELL,MOD
              INTEGER     NTC,OFFSET,TALLY(26),LENGTH,ANIMAL,BLOCK
              REAL        VERSION,MAGFACTOR,VOL,KIND,GRID,LABEL(20),LTO,PT


      C   constants:
              VERSION=1.006    ! program version number.
              OFFSET=64        ! offset of character with decimal. value.
              BELL=7           ! bell tone for terminal.

      C   output title, open disk file, determine data kind:
              WRITE(7,100) VERSION
 100          FORMAT(1H1,T14,'Stereology Data Entry',
           *       ' Program    Version ',F5.3///)
              WRITE (7,110)
 110          FORMAT ( ' PROGRAM LIMITS---',//,'  # of Animals/Group:  15',/,
           *          '  # of Blocks/Animal:  20',/,'  # of Fields/Block:',
           *          '  60',//)
              WRITE(7,120)
 120          FORMAT(3X,'Input name of new disk file')
              READ(5,130) DSKFILE
 130          FORMAT(11A1)
              DSKFILE(11)=0      ! filename ends with ASCII null.
              OPEN( UNIT=4, NAME=DSKFILE, TYPE='NEW', RECORDSIZE=80 )
 135          WRITE(7,140)
 140          FORMAT(3X,'Enter data kind (LMO,LM2,orEM1)')
              READ(5,150) KIND
 150          FORMAT(A4)
              IF( KIND.EQ.'LMO ' ) GOTO 160
              IF( KIND.EQ.'LM2 ' ) GOTO 170
              IF( KIND.EQ.'EM1 ' ) GOTO 175
              GOTO 135                          ! undefined data kind.
      C   define tally characters for LMO data:
 160          CHAR(1)='A'       ! Ppr, point on parenchyma.
              CHAR(2)='S'       ! Pnp, point on nonparenchyma.
              NTC=2             ! number of tally characters.
              LABEL(1)='Ppr '
              LABEL(2)='Pnp '
              GOTO 200

      C   define tally characters for LM2 data:
 170          CHAR(1)='A'       ! Pa, Point on Alveolar lumen.
              CHAR(2)='S'       ! Pad, "     "  Alveolar duct lumen.
              CHAR(3)='D'       ! Pt
              CHAR(4)='Q'       ! Pvl
              CHAR(5)='W'       ! Pvw
              CHAR(6)='E'       ! Pal
              CHAR(7)='R'       ! Paw
              CHAR(8)='F'       ! Pbrl
```

```
            CHAR(9)='G'        ! Pbrw
            CHAR(10)='Z'       ! Pbl
            CHAR(11)='X'       ! Pbw
            CHAR(12)='C'       ! Prbl
            CHAR(13)='V'       ! Prbw
            CHAR(14)='B'       ! Pfc
            CHAR(15)='L'       ! It
            CHAR(16)='P'       ! Iaw
            CHAR(17)='O'       ! Ivw
            CHAR(18)='M'       ! Ibw
            CHAR(19)='N'       ! Irbw
            LABEL(1)='Pa  '
            LABEL(2)='Pad '
            LABEL(3)='Pt  '
            LABEL(4)='Pvl '
            LABEL(5)='Pvw '
            LABEL(6)='Pal '
            LABEL(7)='Paw '
            LABEL(8)='Pbrl'
            LABEL(9)='Pbrw'
            LABEL(10)='Pbw '
            LABEL(11)='Pbl '
            LABEL(12)='Prbl'
            LABEL(13)='Prbw'
            LABEL(14)='Pfc '
            LABEL(15)='Pt  '
            LABEL(16)='Iaw '
            LABEL(17)='Ivw '
            LABEL(18)='Ibw '
            LABEL(19)='Irbw'
            NTC=19             ! number of tally characters.
            GOTO 200

C     define tally characters for EM1 data:
175         CHAR(1)='A'        ! Pa
            CHAR(2)='S'        ! Pep1
            CHAR(3)='Z'        ! Pep2
            CHAR(4)='W'        ! Pcen
            CHAR(5)='D'        ! Pcl
            CHAR(6)='E'        ! Pbi
            CHAR(7)='Q'        ! Psi
            CHAR(8)='C'        ! Ps
            CHAR(9)='X'        ! Pam
            CHAR(10)='L'       ! Iep1
            CHAR(11)='P'       ! Iep2
            CHAR(12)='O'       ! Ienw
            LABEL(1)='Pa  '
            LABEL(2)='Pep1'
            LABEL(3)='Pep2'
            LABEL(4)='Pcen'
            LABEL(5)='Pcl '
            LABEL(6)='Pbi '
            LABEL(7)='Psi '
            LABEL(8)='Ps  '
            LABEL(9)='Pam '
            LABEL(10)='Iep1'
            LABEL(11)='Iep2'
            LABEL(12)='Ienw'
            NTC=12             ! number of tally characters.
            GOTO 200

C     input labeling and comments for top of data file:
C           file title:
```

```
200       WRITE(7,210)
210       FORMAT(3X,'Input file title')
          READ(5,220) BUF1
220       FORMAT(60A1)
          WRITE(4,230) BUF1
230       FORMAT('T  Title',T17,':',60A1)

C         investigator:
          WRITE(7,240)
240       FORMAT(3X,'Input investigator')
          READ(5,220) BUF1
          WRITE(4,245) BUF1
245       FORMAT('T  Investigator :',60A1)

C         date:
          CALL DATE(LDATE(1))
          WRITE(4,247) LDATE
247       FORMAT('T  Date           :',9A1)

C         disk number:
          WRITE(7,250)
250       FORMAT(3X,'Input disk number')
          READ(5,220) BUF1
          WRITE(4,255) BUF1
255       FORMAT('T  Disk no.      :',60A1)

C         file name:
          WRITE(4,260) DSKFILE
260       FORMAT('T  File name     :',60A1)

C         data kind:
          WRITE(4,270) KIND
270       FORMAT('T  Data kind',T17,':',A4)

C  grid type used:
          CALL GRDTYP(GRID,MOD,PT,LTO)
          WRITE(4,275) GRID,MOD,PT,LTO
275       FORMAT('T  Grid used',T17,':',A4,A1,2F10.3)

C         comment lines:
          WRITE(7,280)
280       FORMAT(3X,'Input file comments. End with "//" at start of line')
290       READ(5,300) BUF2
300       FORMAT(77A1)
          IF( (BUF2(1).EQ.'/').AND.(BUF2(2).EQ.'/') ) GOTO 315
          WRITE(4,310) BUF2
310       FORMAT('C  ',77A1)
          GOTO 290

C  page screen and output data set choosen:
315       WRITE(7,100) VERSION
          WRITE(7,320) KIND
320       FORMAT(3X,A4,' Tally set:')
          DO 340 I=1,NTC
              WRITE(7,330) LABEL(I),CHAR(I)
330           FORMAT(6X,A4,' : ',A1)
340       CONTINUE

C     output instructions:
          WRITE (7,345)
345       FORMAT (///' DATA INPUT INSTRUCTIONS',//,' ....Use only',
     *                '  letters which have been defined.',/,
     *                '  ....Place one slash (/) at the end of each field.',
```

```
       *              /,'  ....Or if you are finished with the block,',/,
       *              '        place a double slash (//) at the end',
       *              ' of that field.')
              GO TO 400

C*******READ DATA FILE DATA***********************************************
C     block header:
C         done?
350       WRITE(7,360)
360       FORMAT(3X,'Any more blocks for this animal?(Y/N)')
          READ(5,370) OPTION
370       FORMAT(A1)
          IF( OPTION.EQ.'Y' ) GOTO 455
          WRITE(7,380)
380       FORMAT(3X,'Any more animals from same group to be input(Y/N)')
          READ(5,370) OPTION
          IF( OPTION.EQ.'N' ) GOTO 900

C         block header:
400       WRITE(7,410)
410       FORMAT(/3X,'Input animal ID')
          READ(5,420) ANIMAL
420       FORMAT(I5)
          WRITE(7,421)
421       FORMAT(3X,'Input any animal comments. End with "//" at ',
      *          'start of line')
422       READ(5,300) BUF2
          IF( (BUF2(1).EQ.'/').AND.(BUF2(2).EQ.'/') ) GOTO 435
          WRITE(4,310) BUF2
          GOTO 422
435       WRITE(7,440)
440       FORMAT(3X,'Input lung volume (cc)')
          READ(5,450,ERR=435) VOL
450       FORMAT(F7.1)
455       WRITE(7,457)
457       FORMAT(3X,'Input block ID')
          READ(5,420) BLOCK
          IF (KIND.EQ.'LM0') GO TO 465
460       WRITE(7,462)
462       FORMAT(3X,'Input D line length')
          READ(5,450,ERR=460) MAGFACTOR
          GO TO 469
465       MAGFACTOR=1.0
469       WRITE(4,470) ANIMAL,BLOCK,MAGFACTOR,VOL
470       FORMAT('H Animal:',I5,'   Block:',I5,'    D in mic:',F7.3,
      *          '  Volume(cc):',F7.2)

          DO 480 J=1,26    ! initialize tally array.
480          TALLY(J)=0

          WRITE(7,490)
490       FORMAT(3X,'Start data input..')

          DONE=.FALSE.     ! flag for end of block.
C     input a line of data:
500       READ(5,510) DATA
510       FORMAT(80A1)

C     find line length:
          LENGTH=80
520       IF( DATA(LENGTH).NE.' ' ) GOTO 530
```

```
              LENGTH=LENGTH-1
              IF( LENGTH.EQ.0 ) GOTO 500
              GOTO 520

C     tally the data line:
530           DO 700 I=1,LENGTH
                  DO 540 J=1,NTC
                      IF( DATA(I).NE.CHAR(J) ) GOTO 540
                      K=CHAR(J)-OFFSET
                      TALLY(K)=TALLY(K)+1
                      GOTO 700
540               CONTINUE
                  IF( DATA(I).EQ.' ' ) GOTO 700        ! skip blank lines.
                  IF( DATA(I).EQ.'/' ) GOTO 580        ! end of field
                  WRITE(7,560) BELL,DATA(I)            ! undefined character.
560               FORMAT(3X,A1,'Undefined character ',A1,' found ',
     *                  'and deleted.')
                  GOTO 700
580               IF( DATA(I+1).EQ.'/' ) DONE=.TRUE.
                  WRITE(4,590) (TALLY(CHAR(J)-OFFSET),J=1,NTC)
590               FORMAT('D ',19(I3,1X))
                  DO 600 J=1,26                        ! reset tally array.
600                   TALLY(J)=0

                  IF( DONE ) GOTO 350                  ! end of block check.
                  GOTO 500                             ! read next field.

700           CONTINUE

              GOTO 500

900           CLOSE( UNIT=4 )

              STOP 'File saved... bye...'
              END
```

```
      SUBROUTINE GRDTYP(GRID,MOD,PT,LTO)

C*****************************************************************
C   GRDTYP                                        EFR 11-APR-80   *
C                                                                 *
C   Routine for determining the grid type used in collecting      *
C   the Stereology data for INPUT1 and INPUT2.                    *
C                                                                 *
C*****************************************************************

C  Variables:
      LOGICAL*1 MOD,TYPE(6)
      REAL      GRID,PT,LTO

      DATA TYPE/ 'A','B','C','D','E','F' /
      GRID='   '
      MOD=' '
      ERROR=.FALSE.

C  input grid used:
100   WRITE(7,110)
110   FORMAT(3X,'Input grid type used (A100,B100,C64,D64,M42,M168)')
      READ(5,120) GRID
120   FORMAT(A4)
      IF( GRID.EQ.'A100' ) GOTO 200
      IF( GRID.EQ.'B100' ) GOTO 300
      IF( GRID.EQ.'C64 ' ) GOTO 400
      IF( GRID.EQ.'D64 ' ) GOTO 500
      IF( GRID.EQ.'M42 ' ) GOTO 600
      IF( GRID.EQ.'M168' ) GOTO 700
      WRITE(7,140)
140   FORMAT('    undefined grid type... try again...')
      GOTO 100

C  Grid is A100:
200   WRITE(7,210)
210   FORMAT('   Input line format used for grid:')
      WRITE(7,220)
220   FORMAT('     A: All grid lines used.'/
     *       '     B: Vertical (or horizonal) lines only.')

      CALL GETMOD(MOD,TYPE,2)

C     define LTO:
      PT=100.
      IF(MOD.EQ.'A')  LTO=2.0
      IF(MOD.EQ.'B')  LTO=1.0
      RETURN

C  Grid is B100:
300   WRITE(7,210)
      CALL WRIFOR      ! write legal formats to term.
      CALL GETMOD(MOD,TYPE,6)

C     define LTO:
      PT=100.
      IF( (MOD.EQ.'D').OR.(MOD.EQ.'F') ) PT=400.
      IF(MOD.EQ.'A')  LTO=2.0
      IF(MOD.EQ.'B')  LTO=1.0
      IF(MOD.EQ.'C')  LTO=4.0
      IF(MOD.EQ.'D')  LTO=1.0
      IF(MOD.EQ.'E')  LTO=2.0
      IF(MOD.EQ.'F')  LTO=0.5
```

```
            RETURN

C    Grid is C64:
400       WRITE(7,210)
          CALL WRIFOR
          CALL GETMOD(MOD,TYPE,6)

C         define LT0:
          PT=64.
          IF( (MOD.EQ.'D').OR.(MOD.EQ.'F') ) PT=576.
          IF(MOD.EQ.'A')   LT0=2.0
          IF(MOD.EQ.'B')   LT0=1.0
          IF(MOD.EQ.'C')   LT0=6.0
          IF(MOD.EQ.'D')   LT0=2.0/3.0
          IF(MOD.EQ.'E')   LT0=3.0
          IF(MOD.EQ.'F')   LT0=1.0/3.0
          RETURN

C    Grid is D64:
500       WRITE(7,210)
          CALL WRIFOR
          CALL GETMOD(MOD,TYPE,6)

C         define LT0:
          PT=64.
          IF( (MOD.EQ.'D').OR.(MOD.EQ.'F') ) PT=1024.
          IF(MOD.EQ.'A')   LT0=2.0
          IF(MOD.EQ.'B')   LT0=1.0
          IF(MOD.EQ.'C')   LT0=8.0
          IF(MOD.EQ.'D')   LT0=0.5
          IF(MOD.EQ.'E')   LT0=4.0
          IF(MOD.EQ.'F')   LT0=0.25
          RETURN

C    grid is M42:
600       PT=42.
          LT0=0.5
          RETURN

C    grid is M168:
700       PT=168.
          LT0=0.5
          RETURN

          END


          SUBROUTINE GETMOD(MOD,TYPE,N)
C***********************************************************************
C                                                                      *
C    This subroutine reads a single character and puts it in           *
C    LOGICAL*1 variable MOD.                                           *
C    If the character read is not in the set of the first N            *
C    characters of TYPE array (LOGICAL*1 array holding list of         *
C    legal characters) an error message is output and a new            *
C    character is read.                                                *
C                                                                      *
C***********************************************************************
          LOGICAL*1 MOD,TYPE(6),BUF(72),ERROR
          INTEGER   N
```

```
100     READ(5,120) BUF                 ! input character.
120     FORMAT(72A1)
        CALL FIND(MOD,BUF,72)           ! find the character input.
        CALL CHECK(MOD,TYPE,N,ERROR)    ! is it a legal character?
        IF(.NOT.ERROR) RETURN           ! exit point.
        WRITE(7,140)
140     FORMAT('      illegal entry... try again...')
        GOTO 100
        END


        SUBROUTINE FIND(MOD,BUF,N)
C*****************************************************************
C                                                                *
C   This routine searches N-character BUF for the first          *
C   non-blank character and puts it in MOD.                      *
C   If no non-blank characters are found in BUF then MOD         *
C   returns with "blank" value.                                  *
C                                                                *
C*****************************************************************

        LOGICAL*1 MOD,BUF(N)

        MOD=' '
        DO 200 I=1,N
           IF( BUF(I).EQ.' ' ) GOTO 200
           MOD=BUF(I)
           RETURN
200     CONTINUE

        RETURN
        END

        SUBROUTINE CHECK(MOD,TYPE,N,ERROR)
C*****************************************************************
C                                                                *
C   This routine checks the first N elements of array TYPE for  *
C   the occorance of character MOD. If MOD is not in the legal   *
C   set (array TYPE) than the error flag is set to true.         *
C                                                                *
C*****************************************************************

        LOGICAL*1 MOD,TYPE(N),ERROR

        ERROR=.FALSE.
        DO 100 I=1,N
           IF( MOD.EQ.TYPE(I) ) RETURN
100     CONTINUE
        ERROR=.TRUE.
        RETURN
        END


        SUBROUTINE WRIFOR
C*****************************************************************
C                                                                *
C   This routine lists the legal format options for grids:       *
C   B100, C64, and D64 on the terminal.                          *
C                                                                *
C*****************************************************************
```

```
      WRITE(7,100)
100   FORMAT(5X,'A: Course lines only, both horizonal and vertical'/
     *      5X,'B: Course lines only, one direction only'/
     *      5X,'C: All lines used, both directions, course points'/
     *      5X,'D: All lines used, both directions, all points'/
     *      5X,'E: All lines used, one direction only,course points'/
     *      5X,'F: All lines used, one direction only, all points')
      RETURN
      END
```

Stereology Data Entry Program IN2

```
      PROGRAM IN2

C**************************************************************************
C  IN2.FOR        Stereology Data Entry Program.        EFR 11/mar/80    *
C                                                       NKT 16-OCT-80    *
C                                                                        *
C    This program is used for entering stereology data that has          *
C  already been tallied.                                                 *
C                                                                        *
C    Variable definitions:                                               *
C                                                                        *
C        BIDN    : Block ID number (integer).                            *
C        BUF1    : Sixty character buffer array.                         *
C        BUF2    : Seventytwo character buffer array.                    *
C        BUF3    : Seven character buffer array.                         *
C        D       : D-line length in microns (real, from screen used).    *
C        DATA    : Twenty element array holds tallied data for a         *
C                  field (integer).                                      *
C        DKIND   : Data type (kind) to be input.                         *
C        ERROR   : Error flag (logical*1).                               *
C        ID      : Animal identification number (integer).               *
C        LABEL   : Twenty element array holds labels for identification  *
C                  of the tallies entered (real).                        *
C        LDATE   : Today's date (real array).                            *
C        LIST    : List of animals (by ID) in OLDFILE.                   *
C        LTO     : Total line length factor.                             *
C        MARKER  : Column marker for input of tallies.                   *
C        NANML   : Number of animals.                                    *
C        NBLKS   : Number of blocks.                                     *
C        NEWFILE : New file name (file to be created).                   *
C        NFIELD  : Number of fields.                                     *
C        NW      : Number of animals written to OLDFILE.                 *
C        OLDFILE : Name of file to be added to.                          *
C        OPTION  : Single character response.                            *
C        PT      : Point total.                                          *
C        VERSION : Program version.                                      *
C        VOL     : Displaced lung volume in cc.                          *
C        WLIST   : List of animals written to NEWFILE. If animal is on   *
C                  LIST and not WLIST it still has to be written to      *
C                  NEWFILE.                                              *
C                                                                        *
C  Object files needed when linking IN2:                                 *
C         ( to link IN2 use LNKIN2.COM )                                 *
C                                                                        *
C         IN2                                                            *
C           GRDTYP                                                       *
C           CONVRT                                                       *
C           TRANS                                                        *
C           CKNAME                                                       *
C           MAKLST                                                       *
C           MOVAN                                                        *
C           MOVTOP                                                       *
C           RDANML                                                       *
C           DEFTYP                                                       *
C           FLINFO                                                       *
C           RTSD                                                         *
C                                                                        *
C**************************************************************************

C  variables:
       LOGICAL*1    NEWFILE(11),LDATE(9),BUF1(60),BUF2(72),BUF3(7),
      *             OLDFILE(11),OPTION,ERROR
```

Programs for Estimating Stereological Values of Mammalian Lung and Heart

```
              INTEGER       ID,N,NBLKS,NFIELD,BIDN,DATA(20),NANML,LIST(50),
            *               WLIST(50),NW
              REAL          VERSION,DKIND,D,VOL,MARKER,LABEL(20),PT,LTO
C   constants:
              VERSION=2.001     ! program version number.
              MARKER='[ ]'      ! column marker for input of tallies.
              ERROR=.FALSE.     ! initialize error flag.
              NW=0              ! number of animals written to NEWFILE.

C   output program title, determine prog. mode:
              WRITE(7,100) VERSION
100           FORMAT(1H1,T14,'Stereology Data Entry',
            *              ' Program      Version ',F5.3///)
              WRITE (7,110)
110           FORMAT (' PROGRAM LIMITS---',//,'    # of Animals/Group:  15',/,
            *              '    # of Blocks/Animal:  20',/,'    # of Fields/Block:',
            *              '    60',//)

              WRITE(7,120)
120           FORMAT('    Do you want to add to an existing data ',
            *              'file (Y/N)?')
125           READ(5,130) BUF2
130           FORMAT(72A1)
              CALL FIND(OPTION,BUF2,72)
              IF( OPTION.EQ.'Y' ) GOTO 200      ! set response.
              IF( OPTION.EQ.'N' ) GOTO 300      ! add to existing file.
              WRITE(7,140)                      ! create new data file.
140           FORMAT('    What?')
              GOTO 125

C--add to an existing data file------------------------------------
C         read and open old data file:
200           WRITE(7,210)
210           FORMAT('    Input name of old data file')
              READ(5,130) BUF2
              CALL CKNAME(OLDFILE,BUF2,72,ERROR)
              IF(ERROR) GOTO 200
              OPEN( UNIT=3, NAME=OLDFILE, TYPE='OLD', RECORDSIZE=130)

C         read and open new data file:
220           WRITE(7,230)
230           FORMAT('    Input name of new data file')
              READ(5,130) BUF2
              CALL CKNAME(NEWFILE,BUF2,72,ERROR)
              IF(ERROR) GOTO 220
              OPEN( UNIT=4, NAME=NEWFILE, TYPE='NEW', RECORDSIZE=130)

C   create new data file combining old and new animal data:
C         make a list of animals in OLDFILE:
              CALL MAKLST(NANML,LIST)

C         transfer file header information to NEWFILE & define labels:
              CALL MOVTOP(NEWFILE,DKIND,LABEL,LNGTH1,LNGTH2)

C         set ID number for an animal to be added:
240           WRITE(7,250)
250           FORMAT('    Input animal ID')
              READ(5,130) BUF3
              CALL CONVRT(BUF3,7,ID,ERROR)
              IF(ERROR) GOTO 240

C         is there already data for this animal in OLDFILE?
```

```
              DO 260 I=1,NANML
                  IF( ID.EQ.LIST(I) ) CALL MOVAN(ID)    ! if yes: MOVe ANiMaL
260           CONTINUE

C         read new animal data into NEWFILE:
              CALL RDANML(ID,LABEL,LNGTH1,LNGTH2)

C         add animal (ID) to list of animals written to NEWFILE:
              NW=NW+1         ! nw is number of animals written.
              WLIST(NW)=ID

C         is there any more animal data to be added?
270           WRITE(7,280)
280           FORMAT(' Any more animals (Y/N)?')
              READ(5,130) BUF2
                  CALL FIND(OPTION,BUF2,72)
                  IF( OPTION.EQ.'Y' ) GOTO 240
                  IF( OPTION.EQ.'N' ) GOTO 290
                  WRITE(7,140)
                  GOTO 270

C         write any left over data from OLDFILE to NEWFILE:
290           DO 295 I=1,NANML

C             has animal been written to NEWFILE?
              DO 293 J=1,NW
                  IF( LIST(I).EQ.WLIST(J) ) GOTO 295
293           CONTINUE

C             no he hasn't, write him from OLDFILE to NEWFILE:
              CALL MOVAN(LIST(I))

C             yes he has been written to NEWFILE:
295           CONTINUE

297           CONTINUE

              CLOSE( UNIT=3 )
              CLOSE( UNIT=4 )
              STOP 'Data saved... All done...'
C--create a new data file------------------------------------------
300           WRITE(7,320)
320           FORMAT(3X,'Input name of new disk file')
              READ(5,330) BUF2
330           FORMAT(72A1)
                  CALL CKNAME(NEWFILE,BUF2,72,ERROR)
                  IF( ERROR ) GOTO 300
                  OPEN( UNIT=4, NAME=NEWFILE, TYPE='NEW', RECORDSIZE=130)

C define the type of data to be processed:
340           WRITE(7,350)
350           FORMAT('   Enter data kind (LM0,LM2,EM1)')
              READ(5,360) DKIND
360           FORMAT(A4)
              CALL DEFTYP(DKIND,LABEL,LNGTH1,LNGTH2,ERROR)
              IF(ERROR) GOTO 340

C input labeling and comments for top of data file:
              CALL FLINFO(DKIND,NEWFILE)

C read tallied stereology data into OUTFILE:
              CALL RTSD(LABEL,LNGTH1,LNGTH2,DKIND)

              CLOSE( UNIT=4 )
              STOP 'Data saved... All done...'
              END
```

Programs for Estimating Stereological Values of Mammalian Lung and Heart 247

```
            SUBROUTINE GRDTYP(GRID,MOD,PT,LTO)

C*********************************************************************
C   GRDTYP                                        EFR 11-APR-80    *
C                                                                  *
C     Routine for determining the grid type used in collecting     *
C     the Stereology data for INPUT1 and INPUT2.                   *
C                                                                  *
C*********************************************************************

C   Variables:
            LOGICAL*1 MOD,TYPE(6)
            REAL      GRID,PT,LTO

            DATA TYPE/ 'A','B','C','D','E','F' /
            GRID='    '
            MOD=' '
            ERROR=.FALSE.

C   input grid used:
100         WRITE(7,110)
110         FORMAT(3X,'Input grid type used (A100,B100,C64,D64,M42,M168)')
            READ(5,120) GRID
120         FORMAT(A4)
            IF( GRID.EQ.'A100' ) GOTO 200
            IF( GRID.EQ.'B100' ) GOTO 300
            IF( GRID.EQ.'C64 ' ) GOTO 400
            IF( GRID.EQ.'D64 ' ) GOTO 500
            IF( GRID.EQ.'M42 ' ) GOTO 600
            IF( GRID.EQ.'M168' ) GOTO 700
            WRITE(7,140)
140         FORMAT('     undefined grid type... try again...')
            GOTO 100

C   Grid is A100:
200         WRITE(7,210)
210         FORMAT(' Input line format used for grid:')
            WRITE(7,220)
220         FORMAT('     A: All grid lines used.'/
     *             '     B: Vertical (or horizonal) lines only.')

            CALL GETMOD(MOD,TYPE,2)

C           define LTO:
            PT=100.
            IF(MOD.EQ.'A') LTO=2.0
            IF(MOD.EQ.'B') LTO=1.0
            RETURN

C   Grid is B100:
300         WRITE(7,210)
            CALL WRIFOR        ! write legal formats to term.
            CALL GETMOD(MOD,TYPE,6)

C           define LTO:
            PT=100.
            IF( (MOD.EQ.'D').OR.(MOD.EQ.'F') ) PT=400.
            IF(MOD.EQ.'A') LTO=2.0
            IF(MOD.EQ.'B') LTO=1.0
            IF(MOD.EQ.'C') LTO=4.0
            IF(MOD.EQ.'D') LTO=1.0
            IF(MOD.EQ.'E') LTO=2.0
            IF(MOD.EQ.'F') LTO=0.5
```

```
C  Grid is C64:
400      WRITE(7,210)
         CALL WRIFOR
         CALL GETMOD(MOD,TYPE,6)

C        define LTO:
         PT=64.
         IF( (MOD.EQ.'D').OR.(MOD.EQ.'F') ) PT=576.
         IF(MOD.EQ.'A')  LTO=2.0
         IF(MOD.EQ.'B')  LTO=1.0
         IF(MOD.EQ.'C')  LTO=6.0
         IF(MOD.EQ.'D')  LTO=2.0/3.0
         IF(MOD.EQ.'E')  LTO=3.0
         IF(MOD.EQ.'F')  LTO=1.0/3.0
         RETURN

C  Grid is D64:
500      WRITE(7,210)
         CALL WRIFOR
         CALL GETMOD(MOD,TYPE,6)

C        define LTO:
         PT=64.
         IF( (MOD.EQ.'D').OR.(MOD.EQ.'F') ) PT=1024.
         IF(MOD.EQ.'A')  LTO=2.0
         IF(MOD.EQ.'B')  LTO=1.0
         IF(MOD.EQ.'C')  LTO=8.0
         IF(MOD.EQ.'D')  LTO=0.5
         IF(MOD.EQ.'E')  LTO=4.0
         IF(MOD.EQ.'F')  LTO=0.25
         RETURN

C  grid is M42:
600      PT=42.
         LTO=0.5
         RETURN

C  grid is M168:
700      PT=168.
         LTO=0.5
         RETURN

         END


         SUBROUTINE GETMOD(MOD,TYPE,N)
C**********************************************************************
C                                                                     *
C    This subroutine reads a single character and puts it in          *
C  LOGICAL*1 variable MOD.                                            *
C    If the character read is not in the set of the first N           *
C  characters of TYPE array (LOGICAL*1 array holding list of          *
C  legal characters) an error message is output and a new             *
C  character is read.                                                 *
C                                                                     *
C**********************************************************************
         LOGICAL*1 MOD,TYPE(6),BUF(72),ERROR
         INTEGER   N
```

```
100     READ(5,120) BUF                    ! input character.
120     FORMAT(72A1)
        CALL FIND(MOD,BUF,72)              ! find the character input.
        CALL CHECK(MOD,TYPE,N,ERROR)       ! is it a legal character?
        IF(.NOT.ERROR) RETURN              ! exit point.
        WRITE(7,140)
140     FORMAT('     illegal entry... try again...')
        GOTO 100
        END


        SUBROUTINE FIND(MOD,BUF,N)
C*****************************************************************
C                                                                 *
C    This routine searches N-character BUF for the first          *
C  non-blank character and puts it in MOD.                        *
C    If no non-blank characters are found in BUF then MOD         *
C  returns with "blank" value.                                    *
C                                                                 *
C*****************************************************************
        LOGICAL*1 MOD,BUF(N)

        MOD=' '
        DO 200 I=1,N
           IF( BUF(I).EQ.' ' ) GOTO 200
           MOD=BUF(I)
           RETURN
200     CONTINUE

        RETURN
        END

        SUBROUTINE CHECK(MOD,TYPE,N,ERROR)
C*****************************************************************
C                                                                 *
C    This routine checks the first N elements of array TYPE for   *
C  the occorance of character MOD. If MOD is not in the legal     *
C  set (array TYPE) than the error flag is set to true.           *
C                                                                 *
C*****************************************************************
        LOGICAL*1 MOD,TYPE(N),ERROR

        ERROR=.FALSE.
        DO 100 I=1,N
           IF( MOD.EQ.TYPE(I) ) RETURN
100     CONTINUE
        ERROR=.TRUE.
        RETURN
        END


        SUBROUTINE WRIFOR
C*****************************************************************
C                                                                 *
C    This routine lists the legal format options for grids:       *
C  B100, C64, and D64 on the terminal.                            *
C                                                                 *
C*****************************************************************
```

```
      WRITE(7,100)
100   FORMAT(5X,'A: Course lines only, both horizonal and vertical'/
     *       5X,'B: Course lines only, one direction only'/
     *       5X,'C: All lines used, both directions, course points'/
     *       5X,'D: All lines used, both directions, all points'/
     *       5X,'E: All lines used, one direction only,course points'/
     *       5X,'F: All lines used, one direction only, all points')
      RETURN
      END
```

```
      SUBROUTINE CONVRT(ALPHA,N,NUMBER,ERROR)
C*********************************************************************
C     ROUTINE CONVRT                          efr 18-Jul-80           *
C                                                                     *
C     Converts an alpha string of digits to an integer.               *
C     The number can be anywhere in the string and imbeded            *
C     blanks are ignored.                                             *
C                                                                     *
C     Parameters (passed by calling routine):                         *
C                                                                     *
C         ALPHA    : Logical*1 array holding the string of ASCII      *
C                    digits to be converted into an integer. Note     *
C                    that ALPHA can be an array element also (so      *
C                    you can convert an ASCII digit string which      *
C                    is embedded within other ASCII characters).      *
C         N        : The length of the digit string (number of        *
C                    characters).                                     *
C                                                                     *
C     Parameters (returned by CONVRT)                                 *
C                                                                     *
C         NUMBER   : The decimal value represented by the digit       *
C                    string in ALPHA.                                 *
C         ERROR    : Syntax error flag.                               *
C                                                                     *
C                                                                     *
C*********************************************************************

C Variables:
      INTEGER      N,NUMBER
      LOGICAL*1    ALPHA(N),FLAG,ERROR

      NUMBER=0              ! reset number.
      FLAG=.FALSE.          ! flag for imbedded blanks.
      ERROR=.FALSE.         ! error status.
      M=1                   ! magnitude multiplier.

C convert string to integer:
      DO 300 I=1,N
         J=N+1-I
         IF( ALPHA(J).EQ.' ' ) GOTO 250

C        which digit is ALPHA(J)?
         DO 150 K=0,9
            IF( (ALPHA(J)-48).EQ.K ) GOTO 200
150      CONTINUE

C        non-digit encountered or number to large:
175      ERROR=.TRUE.
         RETURN

C        build number:
200      IF( (K.GE.3).AND.(NUMBER.GT.2767) ) GOTO 175
         IF( NUMBER.GT.10000 ) GOTO 175
         NUMBER=NUMBER+K*M
         IF( M.LT.10000 ) M=10*M
         GOTO 300

C        possible imbedded blank:
250      FLAG=.TRUE.

300   CONTINUE
```

```
      SUBROUTINE TRANS(BUF,N1,DATA,N2,I,J,ISPACE,ERROR)
C*******************************************************************
C   ROUTINE TRANS                                                   *
C                                                                   *
C    Transfers an alpha string of numbers from BUF array to         *
C    DATA array.                                                    *
C       BUF   : LOGICAL*1 array holding a string of ASCII           *
C               digits to be converted into contigous integers      *
C               and stored in DATA array.                           *
C       N1    : INTEGER variable holding the number of elements     *
C               in BUF array.                                       *
C       DATA  : INTEGER target array for the converted numbers      *
C       N2    : INTEGER variable holds the length of DATA array.    *
C       I     : INTEGER, starting location in DATA array for        *
C               the converted numbers.                              *
C       J     : INTEGER, ending location in data array for the      *
C               converted numbers for BUF array.                    *
C       ISPACE: INTEGER, spacing of the integer fields where        *
C               the ASCII numbers will be found in BUF array.       *
C       ERROR : LOGICAL*1 error flag is set when an invalid         *
C               character is encountered during the conversion      *
C               of ASCII characters to integer numbers. Only        *
C               digits (0-9) are valid, no negative numbers.        *
C                                                                   *
C    This routine was written so that streams of integer            *
C    numbers could be input without having to be right              *
C    justified.                                                     *
C                                                                   *
C*******************************************************************
C  variables:
        LOGICAL*1 ALPHA(10),BUF(N1),ERROR
        INTEGER   ISPACE,DATA(N2)

C  do start and stop indices (I,J) for DATA array make sense?
        IF( (J-I).LE.0 ) STOP'error in TRANS...invalid I,J parameters...'
C  is there room in BUF array for J-I+1 numbers?
        IF( (ISPACE*(J-I+1)).LE.N1 ) GOTO 50
        STOP'error in trans sub...BUF array to small...'
C  is I or J out of bounds of DATA array?
50      IF( I.GT.N1 ) STOP'error in trans sub...I out bounds...'
        IF( J.GT.N1 ) STOP'error in trans sub...J out of bounds...'

C  convert string if ASCII characters into discrete numbers:
        DO 300 K=1,(J-I+1)
          DO 100 L=1,ISPACE
100         ALPHA(L)=BUF( (K-1)*ISPACE+L )
          CALL CONVRT(ALPHA,5,DATA(K+I-1),ERROR)
          IF(.NOT.ERROR) GOTO 300
          WRITE(7,200)
200       FORMAT(3X,'Error encountered. Try again...')
          RETURN
300     CONTINUE

        RETURN
        END
```

```
      SUBROUTINE CKNAME(FILE,BUF,N,ERROR)
C*****************************************************************
C  Routine CKNAME                                EFR 16-may-80   *
C                                                                *
C    This subroutine ckecks RT-11 filenames for proper format.   *
C  The RT-11 filenames considered do not have the device kind    *
C  listed, only file name and file type. The following format    *
C  is allowed:                                                   *
C                                                                *
C      name.typ                                                  *
C                                                                *
C  where:      name   : one to six characters. Valid chars are   *
C                       upper case letters (A-Z) or digits       *
C                       (0-9).                                   *
C                     : name/type seperator. There must be a     *
C                       period between the filename and file-    *
C                       type.                                    *
C                 typ : zero to three characters defining the    *
C                       filetype. Valid chars are upper case     *
C                       letters or digits.                       *
C                                                                *
C    Soubroutine LEGAL is called to check the validity of each   *
C  character in the filename. If a filename is found invalid     *
C  CKNAME returns with ERROR set to true.                        *
C                                                                *
C*****************************************************************
        LOGICAL*1    FILE(11),ERROR,BUF(N),ILLEGAL,PERIOD
        INTEGER      ISTART,IEND,NP,N

        NP=0             ! number of periods encountered.
        FILE(11)=0       ! file name ends with ASCII null.
        PERIOD=.FALSE.   ! initialise period flag.
        ERROR=.FALSE.    ! syntax error flag.

C  initialize FILE to blanks:
        DO 50 I=1,10
50         FILE(I)=' '

C  determine start and end points of filename in BUF:
        DO 100 I=1,N
           IF( BUF(I).NE.' ' ) GOTO 110      ! check for character.
100     CONTINUE
        ERROR=.TRUE.                          ! empty buffer
        RETURN

110     ISTART=I
        DO 120 I=1,N
           J=N+1-I
           IF( BUF(J).NE.' ' ) GOTO 130      ! check for character.
120     CONTINUE

130     IEND=J
        IF( (IEND-ISTART).GT.9 ) GOTO 200    ! check filename length.

C  transfer filename from BUF to FILE:
        J=0
        DO 180 I=ISTART,IEND
           CALL LEGAL(BUF(I),ILLEGAL,PERIOD,NP)
           IF( ILLEGAL ) GOTO 200            ! illegal char found.
           IF( PERIOD.AND.(I.EQ.ISTART) ) GOTO 200
```

```
              IF( PERIOD.AND.(I.GT.(ISTART+6)) ) GOTO 200
              IF( PERIOD.AND.((IEND-I).GT.3) ) GOTO 200
              J=J+1
              FILE(J)=BUF(I)
180      CONTINUE
         RETURN               ! normal termination.

C  syntax error encountered:
200      WRITE(7,220)
220      FORMAT('     Syntax error')
         ERROR=.TRUE.
         RETURN               ! return with syntax error.
         END


         SUBROUTINE LEGAL(CHAR,ILLEGAL,PERIOD,NP)


C***********************************************************************
C  LEGAL                                                               *
C                                                                      *
C    Called by CKNAME to chaek for valid filename characters.          *
C                                                                      *
C    Checks to see that CHAR is a legal character. A legal             *
C  character is: upper case letter (A-Z); digit (0-9);                 *
C  or a period if NP is not greater than one. If an                    *
C  illegal character is found ILLEGAL is set, otherwise                *
C  it stays false.                                                     *
C                                                                      *
C***********************************************************************
         LOGICAL*1    CHAR,ILLEGAL,PERIOD
         INTEGER      NP

         ILLEGAL=.FALSE.
         PERIOD=.FALSE.

C  check if character is a period:
         IF( CHAR.NE.'.') GOTO 100
         PERIOD=.TRUE.
         NP=NP+1
         IF( NP.GT.1 ) ILLEGAL=.TRUE.    ! name can have only one period.
         RETURN                          ! exit for period.

C  check if character is upper case letter (A-Z):
100      DO 120 I=65,90
             IF( CHAR.EQ.I ) RETURN      ! exit for legal letter.
120      CONTINUE

C  check if character is digit (0-9):
         DO 140 I=48,57
             IF( CHAR.EQ.I ) RETURN      ! exit for legal digit.
140      CONTINUE

         ILLEGAL=.TRUE.
         RETURN                          ! exit for illegal character.

         END

         RETURN
         END
```

```
        SUBROUTINE MAKLST(NANML,LIST)
C*****************************************************************
C   MAKLST.FOR                               EFR 2-JUL-80         *
C                                                                 *
C    Read through stereology data file and make a list of all     *
C   the animal ID numbers, then rewind the file to its begining.  *
C                                                                 *
C*****************************************************************
        LOGICAL*1  FIRSTCHAR
        INTEGER    NANML,ID,OLDID,LIST(50)

        OLDID=0
        NANML=0

100     READ(3,120,END=160) FIRSTCHAR
120     FORMAT(A1)
        IF( FIRSTCHAR.NE.'H' ) GOTO 100        ! block header line?

        BACKSPACE 3
        READ(3,140) ID
140     FORMAT(10X,I5)
        IF( ID.EQ.OLDID ) GOTO 100             ! new ID number?
        NANML=NANML+1
        LIST(NANML)=ID
        OLDID=ID
        GOTO 100

160     REWIND 3
        RETURN
        END
```

```fortran
      SUBROUTINE MOVAN(ID)
C*****************************************************************
C   MOVAN.FOR                                    EFR 2-JUL-80    *
C                                                                *
C    Move an animal (all blocks of data) from OLDFILE (file 3)   *
C    to NEWFILE (file 4).Note, moves all contigous blocks of data.*
C                                                                *
C*****************************************************************
          LOGICAL*1  FIRSTCHAR,BUF(100)
          INTEGER    ID,AN

C  position to start of OLDFILE:
          REWIND 3

C  find next block header:
100       READ(3,110,END=200) FIRSTCHAR
110       FORMAT(A1)
              IF( FIRSTCHAR.NE.'H' ) GOTO 100

C  found start of block, is this the start of animal number ID?
115       BACKSPACE 3
          READ(3,120) AN
120       FORMAT(10X,I5)
              IF( AN.NE.ID ) GOTO 100

C  found correct animal, copy block over to NEWFILE:
          BACKSPACE 3
          READ(3,140) BUF            ! read block header.
140       FORMAT(100A1)
          WRITE(4,140) BUF
150       READ(3,140,END=200) BUF  ! read field of data.
              IF( BUF(1).EQ.'H' ) GOTO 115
              WRITE(4,140) BUF
              GOTO 150                        ! go back to read next field.

C  all done:
200       RETURN
          END
```

```
        SUBROUTINE MOVTOP(FILE,DKIND,LABEL,LNGTH1,LNGTH2)
C*******************************************************************
C  MOVTOP.FOR                                      EFR 2-JUL-80     *
C                                                                   *
C    Transfer stereology data file header information and group     *
C    comments from the top of OLDFILE (file 3) to NEWFILE (file 4   *
C    ) and add date of revision.                                    *
C                                                                   *
C*******************************************************************
        LOGICAL*1   MDATE(9),BUF(100),FILE(11),ERROR
        INTEGER     LNGTH1,LNGTH2
        REAL        DKIND,LABEL(20)

        CALL DATE(MDATE(1))             ! get today's date.
C  transfer header information:
        READ(3,120) BUF                 ! title.
        WRITE(4,120) BUF

        READ(3,120) BUF                 ! investigator.
        WRITE(4,120) BUF

        READ(3,140) (BUF(I),I=1,9)      ! data collection date.
        WRITE(4,200) (BUF(I),I=1,9),MDATE

        READ(3,120) BUF                 ! disk number.
        WRITE(4,120) BUF

        READ(3,120)                     ! skip old file name.
        WRITE(4,170) (FILE(I),I=1,10)   ! output new file name.

        READ(3,160) DKIND               ! data kind.
        CALL DEFTYP(DKIND,LABEL,LNGTH1,LNGTH2,ERROR)
        IF( ERROR ) STOP ' Undefined data type encountered. From MOVTOP'
        WRITE(4,180) DKIND

        READ(3,120) BUF                 ! gird,pt,lt0.
        WRITE(4,120) BUF

C  transfer any file comments:
100     READ(3,120) BUF
        IF( BUF(1).EQ.'H' ) GOTO 110
        WRITE(4,120) BUF
        GOTO 100
110     BACKSPACE 3
        RETURN                  ! all done

120     FORMAT(100A1)
140     FORMAT(17X,9A1)
160     FORMAT(17X,A4)
170     FORMAT('T  File name',T17,':',10A1)
180     FORMAT('T  Data kind',T17,':',A4)
200     FORMAT('T  Date',T17,':',9A1,T30,'Revised on:',9A1)

        END
```

```
              SUBROUTINE RDANML(ID,LABEL,LNGTH1,LNGTH2,DKIND)

C*******************************************************************
C   RDANML.FOR                                    EFR 3-JUL-80     *
C                                                                  *
C     Input tallied data for animal number ID.                     *
C                                                                  *
C*******************************************************************
          LOGICAL*1   BUF2(76),BUF3(7),ERROR
          INTEGER     NBLKS,LACT,LNGTH1,LNGTH2,BIDN,NFIELD,DATA(60)
          REAL        MARKER,VOL,LABEL(20),D

          MARKER='[   ]'              ! column marker for input of tallies.
C  input lung vloume:
410       WRITE(7,420)
420       FORMAT(3X,'Input lung volume (cc)')
          READ(5,430,ERR=410) VOL
430       FORMAT(F7.0)

C  input number of blocks:
460       WRITE(7,470) ID
470       FORMAT(3X,'Input number of blocks for animal ',I4)
          READ(5,475,ERR=460) BUF3
475       FORMAT(7A1)
          CALL CONVRT(BUF3,7,NBLKS,ERROR)
          IF( ERROR ) GOTO 460
          IF( (NBLKS.GT.20).OR.(NBLKS.LT.1) ) GOTO 460

C  input block data for the animal:
          DO 640 J=1,NBLKS
            LACT='th'
            IF( J.EQ.1 ) LACT='st'
            IF( J.EQ.2 ) LACT='nd'
            IF( J.EQ.3 ) LACT='rd'
480         WRITE(7,490) J,LACT,ID
490         FORMAT(3X,'Input',I3,A2,' block ID number for animal',I5)
            READ(5,500,ERR=480) BIDN
            IF( BIDN.GT.9999 ) GOTO 480
500         FORMAT(I5)

            IF (DKIND.EQ.'LMO') GO TO 505

C  input test line length D:
501         WRITE(7,502)
502         FORMAT(3X,'Input test line length (mic)')
            READ(5,430,ERR=501) D
            GO TO 510
505         D=1.0

510         WRITE(4,520) ID,BIDN,D,VOL
520         FORMAT('H  Animal:',I5,3X,'Block:',I5,3X,
     *             'D in mic :',F7.3,' Volume(cc):',F7.2)

530         WRITE(7,540)
540         FORMAT(3X,'Input number of fields for this block')
            READ(5,545,ERR=530) BUF3
545         FORMAT(7A1)
            CALL CONVRT(BUF3,7,NFIELD,ERROR)
            IF( ERROR ) GOTO 530
            IF( (NFIELD.GT.60).OR.(NFIELD.LT.1) ) GOTO 530
```

```
C              entering the data for each field:
               DO 600 K=1,NFIELD
                  LACT='th'
                  IF( K.EQ.1 ) LACT='st'
                  IF( K.EQ.2 ) LACT='nd'
                  IF( K.EQ.3 ) LACT='rd'
                  WRITE(7,550) K,LACT,BIDN
550               FORMAT(3X,I2,A2,' field of block ',I4,':'/)

552               WRITE(7,555) (LABEL(L),L=1,LNGTH1)
                  WRITE(7,555) (MARKER,L=1,LNGTH1)
555               FORMAT(15(1X,A4))

557               READ(5,560) BUF2
560               FORMAT(76A1)

C              if blank line reread the line:
                  DO 565 I=1,76
                     IF( BUF2(I).NE.' ' ) GOTO 567
565               CONTINUE
                  GOTO 557

C              transfer data from BUF2 to data array (DATA):
567               CALL TRANS(BUF2,76,DATA,60,1,LNGTH1,5,ERROR)
                  IF(ERROR) GOTO 552

                  IF( LNGTH2.EQ.0 ) GOTO 580
570               WRITE(7,555) (LABEL(L),L=16,15+LNGTH2)
                  WRITE(7,555) ( MARKER ,L=16,15+LNGTH2)
                  READ(5,560) BUF2
                  CALL TRANS(BUF2,76,DATA,60,16,15+LNGTH2,5,ERROR)
                  IF(ERROR) GOTO 570

580               LNGTH=LNGTH1+LNGTH2
                  WRITE(4,590) (DATA(L),L=1,LNGTH)
590               FORMAT('D ',15(I3,1X))
600            CONTINUE
640         CONTINUE

            RETURN
            END
```

```fortran
      SUBROUTINE DEFTYP(DKIND,LABEL,LNGTH1,LNGTH2,ERROR)

C*****************************************************************
C     DEFTYP.FOR                                EFR 3-JUL-80      *
C                                               NKT 02-JUL-81     *
C                                                                 *
C     This subroutine, given the data type (DKIND), defines       *
C     the label set and number of tally variables for each type   *
C     of data.  If DKIND is undefined then the routine returns    *
C     with ERROR set to true.                                     *
C                                                                 *
C         Presently defined data types:                           *
C                                                                 *
C             LM0, LM2, EM1.                                      *
C                                                                 *
C*****************************************************************

          LOGICAL*1   ERROR
          INTEGER     LNGTH1,LNGTH2
          REAL        DKIND,LABEL(20)

          ERROR=.FALSE.                   ! initialize error flag

C     determine data kind (type):
          IF( DKIND.EQ.'LM0 ' ) GOTO 160
          IF( DKIND.EQ.'LM2 ' ) GOTO 170
          IF( DKIND.EQ.'EM1 ' ) GOTO 180
          ERROR=.TRUE.                    ! undefined data type.
          WRITE(7,100)
100       FORMAT(' Undefined data kind. From DEFTYP.')
          RETURN

C     define label set for LM0 data:
160       LABEL(1)='Ppr '                 ! parenchyma
          LABEL(2)='Pnp '                 ! nonparenchyma
          LNGTH1=2
          LNGTH2=0
          RETURN

C     define label set for LM2 data:
170       LABEL(1)='Pa  '
          LABEL(2)='Pad '
          LABEL(3)='Pt  '
          LABEL(4)='Pvlu'
          LABEL(5)='Pvw '
          LABEL(6)='Palu'
          LABEL(7)='Paw '
          LABEL(8)='Pbrl'
          LABEL(9)='Pbrw'
          LABEL(10)='Pbl '
          LABEL(11)='Pbw '
          LABEL(12)='Prbl'
          LABEL(13)='Prbw'
          LABEL(14)='Pfc '
          LABEL(15)='It  '
          LABEL(16)='Iaw '
          LABEL(17)='Ivw '
          LABEL(18)='Ibw '
          LABEL(19)='Irbw'
          LNGTH1=15
          LNGTH2=4
          RETURN
```

```
C    define label set for EM1 data:
180      LABEL(1)='Pa   '
         LABEL(2)='Pep1'
         LABEL(3)='Pep2'
         LABEL(4)='Pcen'
         LABEL(5)='Pcl  '
         LABEL(6)='Pbi  '
         LABEL(7)='Psi  '
         LABEL(8)='Ps   '
         LABEL(9)='Pam  '
         LABEL(10)='Iep1'
         LABEL(11)='Iep2'
         LABEL(12)='Ienw'
         LNGTH1=12
         LNGTH2=0
         RETURN

         END
```

A Guide to Practical Stereology

```
      SUBROUTINE FLINFO(DKIND,NEWFILE)
C*********************************************************************
C     FLINFO.FOR                              EFR 3-JUL-80        *
C                                                                 *
C     Create top of stereology data file including group          *
C     comments.                                                   *
C                                                                 *
C*********************************************************************
      LOGICAL*1   LDATE(9),BUF1(60),BUF2(76),NEWFILE(11),MOD
      REAL        DKIND,GRID,PT,LTO

      PT=0.0           ! initialize PT,LTO
      LTO=0.0

C     file title:
200   WRITE(7,210)
210   FORMAT(3X,'Input file title')
      READ(5,220) BUF1
220   FORMAT(60A1)
      WRITE(4,230) BUF1
230   FORMAT('T  Title',T17,':',60A1)

C     investigator:
      WRITE(7,240)
240   FORMAT(3X,'Input investigator')
      READ(5,220) BUF1
      WRITE(4,245) BUF1
245   FORMAT('T  Investigator :',60A1)

C     date:
      CALL DATE(LDATE(1))
      WRITE(4,247) LDATE
247   FORMAT('T  Date         :',9A1)

C     disk number:
      WRITE(7,250)
250   FORMAT(3X,'Input disk number')
      READ(5,220) BUF1
      WRITE(4,255) BUF1
255   FORMAT('T  Disk no.     :',60A1)

C     file name:
      WRITE(4,260) NEWFILE
260   FORMAT('T  File name    :',60A1)

C     data kind:
      WRITE(4,270) DKIND
270   FORMAT('T  Data kind',T17,':',A4)

C     grid type used:
      CALL GRDTYP(GRID,MOD,PT,LTO)
      WRITE(4,275) GRID,MOD,PT,LTO
275   FORMAT('T  Grid used',T17,':',A4,A1,2F10.3)

C     comment lines:
      WRITE(7,280)
280   FORMAT(3X,'Input file comments. End with "//" at start of line')
290   READ(5,300) BUF2
300   FORMAT(77A1)
      IF( (BUF2(1).EQ.'/').AND.(BUF2(2).EQ.'/') ) RETURN
      WRITE(4,310) BUF2
310   FORMAT('C   ',77A1)
      GOTO 290

      END
```

```
            SUBROUTINE RTSD(LABEL,LNGTH1,LNGTH2,DKIND)
C*******************************************************************
C   RTSD.FOR                                       EFR 3-JUL-80     *
C                                                                   *
C     Input animal ID and read tallied block data.                  *
C                                                                   *
C*******************************************************************
            LOGICAL*1   BUF3(7),ERROR
            INTEGER     LACT,ID,LNGTH1,LNGTH2
            REAL        LABEL(20)

350         WRITE(7,370)
370         FORMAT(3X,'Input number of animals in group')
            READ(5,380) BUF3
380         FORMAT(7A1)
            CALL CONVRT(BUF3,7,NANML,ERROR)
            IF( ERROR ) GOTO 350
            IF( NANML.GT.40 ) GOTO 350

            DO 700 I=1,NANML
              LACT='th'
              IF( I.EQ.1 ) LACT='st'
              IF( I.EQ.2 ) LACT='nd'
              IF( I.EQ.3 ) LACT='rd'
385           WRITE(7,390) I,LACT
390           FORMAT(3X,'Input ID number for ',I2,A2,' animal')
              READ(5,400,ERR=385) BUF3
400           FORMAT(7A1)

C             if stop was typed then forget the rest of the data:
              DO 500 J=1,7
                  IF( BUF3(J).EQ.'S' ) RETURN
500           CONTINUE

C             convert ascii ID to a number:
              CALL CONVRT(BUF3,7,ID,ERROR)
              IF( ERROR ) GOTO 385
              IF( (ID.GT.9999).OR.(ID.LT.1) ) GOTO 385

              CALL RDANML(ID,LABEL,LNGTH1,LNGTH2,DKIND)

700         CONTINUE

            CLOSE( UNIT=4 )
            STOP 'Data saved... All done...'
            END
```

Calculation of Stereological Data Program CALC

```
C***************************************************************
C     CALC.FOR   Stereology Data Processing Program   EFR 12-may-80  *
C                                                     NKT 22-JUL-81  *
C                                                                    *
C       This program is used to process tallied Stereology data      *
C     files as created by one of the data input programs IN1 or      *
C     IN2.                                                            *
C       This program will process the following data types:          *
C                                                                    *
C         LM0 : light microscope, par/nonpar data.                   *
C         LM2 : hi mag light microscope data                         *
C         EM1 : electron microscope-celluar.                         *
C                                                                    *
C       CALC is linked to the following subroutines:                 *
C                                                                    *
C         CKNAMC                                                     *
C         TOPFL                                                      *
C         RBLOCK                                                     *
C         CBLOCK                                                     *
C         ADDGRP                                                     *
C         GRPRES                                                     *
C         LM0                                                        *
C         LM2                                                        *
C         EM1                                                        *
C                                                                    *
C***************************************************************


C     variables:
          LOGICAL*1    INFILE(11),OUTFILE(11),TITLE(60),
      *                MOD,ERROR,DONE,BUF(80)
          INTEGER      N,NBLKS,AN,BN,NF,OLDAN,B1,B2,NL
          REAL         VERSION,GRID,DKIND,PT,LT0,LT,
      *                VOL,D,VPR
          REAL*8       L(30)

          COMMON       A(20,32),B(61,30),G(18,33)

          DATA L/30*'           '/


          VERSION=2.001              ! program version.
          N=0                        ! number of animals processed.
          NBLKS=0                    ! initialize block counter.
          OLDAN=0                    ! last animal processed.

C     initialize block array (B), animal array (A), and group
C     results array (G):
          DO 20 I=1,61
             DO 30 J=1,30
                B(I,J)=0.0
30           CONTINUE
20        CONTINUE

          DO 40 I=1,20
             DO 50 J=1,32
                A(I,J)=0.0
50           CONTINUE
40        CONTINUE

          DO 60 I=1,18
             DO 70 J=1,33
                G(I,J)=0.0
```

```
70          CONTINUE
60          CONTINUE
C  output program title, input data file and results file names:
            WRITE(7,100) VERSION
100         FORMAT(1H1,T12,'STEREOLOGY DATA PROCESSING PROGRAM
     *              'VERSION ',F5.3)
110         WRITE(7,120)
120         FORMAT(////' Input data file name')
            READ(5,130) BUF
130         FORMAT(80A1)
            CALL CKNAMC(INFILE,BUF,80,ERROR)       ! ckeck syntax of file name.
            IF( ERROR ) GOTO 110

140         WRITE(7,150)
150         FORMAT(' Input group results file name')
            READ(5,130) BUF
            CALL CKNAMC(OUTFILE,BUF,80,ERROR)      ! check syntax of file name.
            IF( ERROR ) GOTO 140
               OPEN( UNIT=3, NAME=INFILE, TYPE='OLD', RECORDSIZE=130)
               OPEN( UNIT=4, NAME=OUTFILE,TYPE='NEW', RECORDSIZE=130)

C  read datafile title information and output results header:
            CALL TOPFL(DKIND,GRID,MOD,PT,LTO,INFILE,OUTFILE,VERSION,ERROR,
     *          TITLE)
            IF(ERROR) STOP' ERROR ENCOUNTERED WHILE READING DATA FILE HEADER'

C  Read and process data for group:
C     input a block of data:
200         CALL RBLOCK(AN,BN,D,VOL,NF,DONE)
            IF( NF.EQ.0 ) GOTO 270

C   check to see if new animal:
            IF (OLDAN.EQ.AN) GO TO 205

C   enter volume fraction of parenchyma if appropriate:
            IF (DKIND.EQ.'LMO') GO TO 205
            WRITE (7,201) AN
201         FORMAT (' Enter volume fraction of parenchymal airspace',
     *          ' for animal #: ',I5)
            READ (5,202) VPR
202         FORMAT (F5.4)

C      have we come to the end of the data file?
205         IF( DONE ) GOTO 220

C      is the block just read part of a new animal?
            IF( (OLDAN.NE.AN).AND.(OLDAN.NE.0) ) GOTO 210

C      total the block and process it:
            CALL CBLOCK(AN,BN,D,VOL,NF,LTO,DKIND,L,NL,VPR,NBLKS)
            OLDAN=AN
            GOTO 200

C      is a new animal. add previous animal results to group
C      results. then start new animal results:
210         CALL ADDGRP(OLDAN,NBLKS,N)
            CALL CBLOCK(AN,BN,D,VOL,NF,LTO,DKIND,L,NL,VPR,NBLKS)
            OLDAN=AN
            GOTO 200

C      done with data file. was last block a new animal?
220         IF( (OLDAN.NE.AN).AND.(OLDAN.NE.0) ) GOTO 230
```

```
C           add last block to last animal results, then add last
C           animal to group results:
            CALL CBLOCK(AN,BN,D,VOL,NF,LTO,DKIND,L,NL,VPR,NBLKS)
            CALL ADDGRP(AN,NBLKS,N)
            GO TO 300

C           end of data file, and last block is a new animal. Add
C           previous animal results to group results array, then
C           add last animal into group array:
230         CALL ADDGRP(OLDAN,NBLKS,N)
            CALL CBLOCK(AN,BN,D,VOL,NF,LTO,DKIND,L,NL,VPR,NBLKS)
            CALL ADDGRP(AN,NBLKS,N)
            GO TO 300

C  error, block read containing no fields:
270         WRITE(7,290) AN,BN
290         FORMAT('     Block encountered which contains no fields (NF=0)'/
     *             '     problem in animal #',I5,'  block #',I5)
            STOP ' from routine CALC'

C  output group results to OUTFILE:
300         CALL GRPRES(N,L,NL,TITLE)

            CLOSE( UNIT=3 )
            CLOSE( UNIT=4 )
            STOP ' ALL DONE...'
            END
```

```
      SUBROUTINE CKNAMC(FILE,BUF,N,ERROR)
C***********************************************************************
C  Routine CKNAMC                                    EFR 16-may-80    *
C                                                                      *
C  This subroutine checks RT-11 filenames for proper format.           *
C  The RT-11 filenames considered do not have the device kind          *
C  listed, only file name and file type. The following format          *
C  is allowed:                                                         *
C                                                                      *
C      name.typ                                                        *
C                                                                      *
C  where:      name    : one to six characters. Valid chars are        *
C                        upper case letters (A-Z) or digits            *
C                        (0-9).                                        *
C              .       : name/type seperator. There must be a          *
C                        period between the filename and file-         *
C                        type.                                         *
C              typ     : zero to three characters defining the         *
C                        filetype. Valid chars are upper case          *
C                        letters or digits.                            *
C                                                                      *
C  Soubroutine LEGAL is called to check the validity of each           *
C  character in the filename. If a filename is found invalid           *
C  CKNAMC returns with ERROR set to true.                              *
C                                                                      *
C***********************************************************************
      LOGICAL*1   FILE(11),ERROR,BUF(N),ILLEGAL,PERIOD
      INTEGER     ISTART,IEND,NP,N

      COMMON      A(20,32),B(61,30),G(18,33)

      NP=0                  ! number of periods encountered.
      FILE(11)=0            ! file name ends with ASCII null.
      PERIOD=.FALSE.        ! initialise period flag.
      ERROR=.FALSE.         ! syntax error flag.
C  initialize FILE to blanks:
      DO 50 I=1,10
50       FILE(I)=' '

C  determine start and end points of filename in BUF:
      DO 100 I=1,N
         IF( BUF(I).NE.' ' ) GOTO 110      ! check for character.
100   CONTINUE
      ERROR=.TRUE.                          ! empty buffer
      RETURN

110   ISTART=I
      DO 120 I=1,N
         J=N+1-I
         IF( BUF(J).NE.' ' ) GOTO 130      ! check for character.
120   CONTINUE

130   IEND=J
      IF( (IEND-ISTART).GT.9 ) GOTO 200    ! check filename length.

C  transfer filename from BUF to FILE:
      J=0
      DO 180 I=ISTART,IEND
         CALL LEGAL(BUF(I),ILLEGAL,PERIOD,NP)
```

```
              IF( ILLEGAL ) GOTO 200              ! illegal char found.
              IF( PERIOD.AND.(I.EQ.ISTART) ) GOTO 200
              IF( PERIOD.AND.(I.GT.(ISTART+6)) ) GOTO 200
              IF( PERIOD.AND.((IEND-I).GT.3) ) GOTO 200
              J=J+1
              FILE(J)=BUF(I)
180       CONTINUE
          RETURN              ! normal termination.
C  syntax error encountered:
200       WRITE(7,220)
220       FORMAT('    Syntax error')
          ERROR=.TRUE.
          RETURN              ! return with syntax error.
          END


          SUBROUTINE LEGAL(CHAR,ILLEGAL,PERIOD,NP)


C***********************************************************************
C  LEGAL                                                               *
C                                                                      *
C     Called by CKNAMC to check for valid filename characters.         *
C                                                                      *
C     Checks to see that CHAR is a legal character. A legal            *
C  character is: upper case letter (A-Z); digit (0-9);                 *
C  or a period if NP is not greater than one. If an                    *
C  illegal character is found ILLEGAL is set, otherwise                *
C  it stays false.                                                     *
C                                                                      *
C***********************************************************************

          LOGICAL*1   CHAR,ILLEGAL,PERIOD
          INTEGER     NP

          COMMON      A(20,32),B(61,30),G(18,33)

          ILLEGAL=.FALSE.
          PERIOD=.FALSE.

C  check if character is a period:
          IF( CHAR.NE.'.' ) GOTO 100
          PERIOD=.TRUE.
          NP=NP+1
          IF( NP.GT.1 ) ILLEGAL=.TRUE.    ! name can have only one period.
          RETURN                          ! exit for period.

C  check if character is upper case letter (A-Z):
100       DO 120 I=65,90
              IF( CHAR.EQ.I ) RETURN      ! exit for legal letter.
120       CONTINUE

C  check if character is digit (0-9):
          DO 140 I=48,57
              IF( CHAR.EQ.I ) RETURN      ! exit for legal digit.
140       CONTINUE

          ILLEGAL=.TRUE.
          RETURN                          ! exit for illegal character.

          END
```

Programs for Estimating Stereological Values of Mammalian Lung and Heart

```
        SUBROUTINE TOPFL(DKIND,GRID,MOD,PT,LTO,INFILE,OUTFILE,
     *                  VERSION,ERROR,TITLE)
C******************************************************************
C   TOPFL.FOR                                      EFR 26-JUN-80   *
C                                                                  *
C   This subroutine is called by CALC.FOR to read the title        *
C   part of the stereology data file and create the output file    *
C   header.                                                        *
C                                                                  *
C******************************************************************
        LOGICAL*1  ERROR,MOD,INFILE(11),OUTFILE(11),LDATE(9),FF,
     *             TITLE(60),INV(60),DCDATE(9)
        REAL       DKIND,GRID,PT,LTO,VERSION
        REAL*8     L(30)

        COMMON     A(20,32),B(61,30),G(18,33)

        ERROR=.FALSE.
        FF=12                          ! form feed character.

        CALL DATE(LDATE(1))            ! get today's date.
        WRITE(4,270) LDATE,VERSION,LDATE

        READ(3,100) TITLE              ! input data file title.

        READ(3,100) INV                ! input investigator.

        READ(3,100) DCDATE             ! input data collection date.

        READ(3,100)                    ! skip disk no.
        READ(3,100)                    ! skip data file name.

        READ(3,110) DKIND              ! input data type.
        READ(3,110) GRID,MOD,PT,LTO    ! input grid specification.

        WRITE(4,200) TITLE,DKIND
        WRITE(4,210) INV,GRID,MOD
        WRITE(4,220) (INFILE(I),I=1,10),PT
        WRITE(4,230) (OUTFILE(I),I=1,10),DCDATE
100     FORMAT(17X,60A1)
110     FORMAT(17X,A4,A1,2F10.3)

200     FORMAT(//     Group results for ',60A1,T87,'Data type: ',A4)
210     FORMAT( '     Investigator ',60A1,T87,'Grid used: ',A4,A1)
220     FORMAT( '     Data on file:    ',10A1,T87,'Grid PT: ',F6.0)
230     FORMAT( '     Output on file: ',10A1,T87,'Data collected on',
     *          1X,9A1/1X,129('-'))
270     FORMAT(3X,9A1,T39,'Stereology Data Processing Program',
     *          ,'     Version ',F5.3,T119,9A1///)

C   skip any file comment to start of first block:
300     READ(3,320) FF
320     FORMAT(A1)
        IF( FF.NE.'H' ) GOTO 300
        BACKSPACE 3
        RETURN                    ! normal exit.

C   eof reached before first block header:
400     WRITE(7,440)
```

```
440     FORMAT('     Eof encountered before first block header.'/
       *            '     Either there is no data in the file or the'/
       *            '     block headers do not start with an "H" in '/
       *            '     column one.')
        STOP ' from subroutine TOPFL'

        END
```

Programs for Estimating Stereological Values of Mammalian Lung and Heart

```
              SUBROUTINE RBLOCK(AN,BN,D,VOL,NF,DONE)
C*****************************************************************
C   RBLOCK.FOR                                   EFR 24-JUN-80   *
C                                                                *
C   This subroutine reads a block of tally data for program      *
C   CALC.FOR to process as stereology data.                      *
C     Data is read from file 3 and stored in B-array.  DONE is   *
C   set true if EOF is encountered while reading a block of      *
C   data.                                                        *
C                                                                *
C        AN    : Animal number (integer).                        *
C        BN    : Block number (integer).                         *
C        D     : D-line length of grid in microns (real).        *
C        VOL   : Lung volume in cc (real).                       *
C        NF    : Number of fields for block (integer).           *
C        DONE  : Flag set true if EOF encountered while reading  *
C                block tallies (logical*1).                      *
C                                                                *
C*****************************************************************
          LOGICAL*1   T,DONE,BUF(72)
          INTEGER     AN,BN,NF
          REAL        D,VOL

          COMMON      A(20,32),B(61,30),G(18,33)

          DONE=.FALSE.     ! initialize eof flag.

C   initialize B-array to zero:
          DO 110 I=1,61
             DO 100 J=1,30
100             B(I,J)=0.0
110       CONTINUE

C   input block header:
          READ(3,120,ERR=180) AN,BN,D,VOL           ! input animal ID info.
120       FORMAT(10X,I5,9X,I5,13X,F7.1,12X,F7.1)

C   input tallies for each field into B-array:
          NF=0
130       READ(3,140,ERR=150,END=170) T,(B(NF+1,J),J=1,20)
140       FORMAT(A1,1X,20(F3.0,1X))
          NF=NF+1
          GOTO 130

C   error set on read. either the next block of data was encountered
C   or there is a format error for that field.
150       BACKSPACE 3
          IF(T.EQ.'H') RETURN       ! normal exit.
          WRITE(7,160) AN,BN,NF+1 ! format error.
160       FORMAT('     Format error on reading field'/
     *           '          Animal #',I5/
     *           '          Block  #',I5/
     *           '          field  #',I5)
          STOP ' from subroutine RBLOCK'

C   eof encountered while reading a field:
170       DONE=.TRUE.
          RETURN

C   error on reading block header:
```

```
180       BACKSPACE 3
          READ(3,190) BUF
190       FORMAT(72A1)
          WRITE(7,200) BUF
200       FORMAT('Format error encountered while reading this line:'/
         *          1X,72A1)
          STOP ' from subroutine RBLOCK'

          END
```

```fortran
      SUBROUTINE CBLOCK(AN,BN,D,VOL,NF,LTO,DKIND,L,NL,VPR
     *          ,NBLKS)
C*******************************************************************
C  CBLOCK.FOR                                   EFR 24-JUN-80   *
C                                               NKT 02-JUL-81   *
C                                                               *
C     Perform data calculations for the block totals in         *
C  the B-array. The particular calculations performed depend    *
C  on the kind of stereology data being processed. There is     *
C  one subroutine for each type of stereological data.          *
C     F is set when the block is processed, so if it wasn't     *
C  set an undefined data kind was encountered and the program   *
C  halts with an error message.                                 *
C                                                               *
C       AN    : Animal number (integer).                        *
C       BN    : Block number (integer).                         *
C       D     : D-line length in microns (real).                *
C       VOL   : Lung volume in cc (real).                       *
C       A     : Array for blocks of one animal.                 *
C       B     : array holds tally data and                      *
C               totals for the block (real).                    *
C       NF    : Number of fields for block (integer).           *
C       DKIND : Data kind (data type to be processed, real).    *
C       NBLKS : Block counter                                   *
C                                                               *
C*******************************************************************
      LOGICAL*1  F
      INTEGER    AN,BN,NF,L,NBLKS
      REAL       D,VOL,LT,LTO,VPR
      REAL*8     L(30)

      COMMON     A(20,32),B(61,30),G(18,33)

      F=.FALSE.          ! initialize process flag.

      IF( DKIND.EQ.'LM0 ' ) CALL LM0(AN,BN,VOL,NF,LTO,
     *                      D,L,NL,F,NBLKS)
      IF( DKIND.EQ.'LM2 ' ) CALL LM2(AN,BN,VOL,NF,LTO,
     *                      D,L,NL,F,NBLKS)
      IF( DKIND.EQ.'EM1 ' ) CALL EM1(AN,BN,VOL,NF,LTO,
     *                      D,L,NL,F,NBLKS,VPR)

      IF(F) RETURN            ! block processed, normal exit.

      WRITE(7,100) DKIND
100   FORMAT('    Undefined data kind [',A4,']')
      STOP ' from subroutine CBLOCK'

      END
```

```fortran
              SUBROUTINE ADDGRP(AN,NBLKS,N)
C**********************************************************************
C  ADDGRP.FOR                                         EFR 26-JUN-80   *
C                                                     NKT 23-JUL-81   *
C                                                                     *
C   This subroutine adds the block data for an animal to the          *
C   group results. The blocks are averaged for the animal and         *
C   added into the group results (G-array).                           *
C                                                                     *
C**********************************************************************
              LOGICAL*1  ERROR
              INTEGER    AN,N,NBLKS
              REAL       AVG(30),SD(30),VAR(30)

              COMMON     A(20,32),B(61,30),G(18,33)

              N=N+1                 ! increment animal counter.

C   calculate averages and standard deviations for the animal:
C     check for divide by zero error:
              IF (NBLKS.GT.0) GO TO 120
                WRITE (7,110) NBLKS
110             FORMAT ('  Divide by zero error called by ADDGRP',/
     *                  '  divisor: number of blocks (NBLKS) =',I5)
                STOP ' from routine ADDGRP'

C   initialize AVG,SD, and VAR:
120           DO 130 I=1,30
                AVG(I)=0.0
                SD(I)=0.0
                VAR(I)=0.0
130           CONTINUE

C     calculations:
C        column averages:
              DO 140 J=1,30
                DO 150 I=1,NBLKS
                  AVG(J)=AVG(J)+A(I,J)
150             CONTINUE
                IF (AVG(J).EQ.0.0) GO TO 140
                AVG(J)=AVG(J)/NBLKS
140           CONTINUE

C   incorporate animal averages into group results:
              DO 200 I=1,30
200             G(N,I)=AVG(I)

              G(N,31)=AN              ! save animal number.
              G(N,32)=NBLKS           ! save number of blocks for animal.
              DO 250 I=1,NBLKS        ! save number of fields for animal.
250             G(N,33)=G(N,33)+A(I,32)

              NBLKS=0                 ! reset block counter (for B-array).

C   rest animal array for next animal (A-array):
              DO 350 I=1,20
                DO 300 J=1,32
300               A(I,J)=0.0
350           CONTINUE

        RETURN
        END
```

```
        SUBROUTINE GRPRES(N,L,NL,TITLE)

C*********************************************************************
C       GRPRES.FOR                              EFR 26-JUN-80     *
C                                               NKT 24-JUL-81     *
C                                                                 *
C   Subroutine for output of group results to OUTFILE. The        *
C   means and SD's are calculated prior to writing the the        *
C   output file. The header part of the output file was already   *
C   written in subroutine TOPFL.                                  *
C                                                                 *
C*********************************************************************

        INTEGER    N,NL
        REAL*8     L(30)
        LOGICAL*1  FF,LDATE(9),TITLE(60)

        COMMON     A(20,32),B(61,30),G(18,33)

        FF=12

        CALL DATE (LDATE(1))


C   calculate means, SD's and variances for all animals of the group:
C       check for divide by zero error:
        IF (N.NE.0) GO TO 20
        WRITE (7,10)
10      FORMAT ('      Divide by zero error in GRPRES')
        STOP     ' from subroutine GRPRES'

C   calculate means, SD and variances:
20      DO 80 J=1,30
            DO 30 I=1,N
                G(N+1,J)=G(N+1,J)+G(I,J)
30          CONTINUE
            IF (G(N+1,J).EQ.0.0) GO TO 80
            G(N+1,J)=G(N+1,J)/N                     ! mean
            IF (N.LE.1) GO TO 80
            DO 40 I=1,N
                G(N+3,J)=G(N+3,J)+(G(I,J)-G(N+1,J))**2
40          CONTINUE
            G(N+3,J)=G(N+3,J)/(N-1)                 ! variance
            G(N+2,J)=SQRT(G(N+3,J))                 ! SD
80      CONTINUE

C   write to output file:
        WRITE(4,100) (IFIX(G(I,31)),I=1,N)
100     FORMAT(/' Animal #   ',11(1X,I5,4X))
        WRITE(4,110) (IFIX(G(I,32)),I=1,N)
110     FORMAT(/' # of blocks',11(2X,I4,4X))

        WRITE(4,120) (IFIX(G(I,33)),I=1,N)
120     FORMAT(/' # of fields',11(2X,I4,4X))

        DO 200 J=1,NL
            WRITE(4,140) L(J),(G(I,J),I=1,N)
140         FORMAT(/1X,A8,3X,11(G10.3))
200     CONTINUE

        WRITE (4,310) FF,LDATE,LDATE
310     FORMAT (A1,3X,9A1,T49,'Stereology Data Processing Program',
```

```
      *           T119,9A1///)
            WRITE(4,320) TITLE
320         FORMAT ('     Group results for ',60A1,//)
            WRITE (4,330)
330         FORMAT (13X,'Mean      ','    SD       ',' VAR  ',/)
            DO 400 J=1,NL
               WRITE (4,340) L(J), (G(I,J),I=N+1,N+3)
340            FORMAT (/,1X,A8,3X,3(G10.3,1X))
400         CONTINUE
            RETURN
            END
```

Programs for Estimating Stereological Values of Mammalian Lung and Heart

```
      SUBROUTINE LMO(AN,BN,VOL,NF,LTO,D,L,NL,FLAG,NBLKS)

C******************************************************************
C     LMO.FOR                                  EFR 24-JUN-80   *
C                                              NKT 24-JUL-81   *
C                                                              *
C     Perform data calculations for LMO stereology data on the *
C     totaled data in B-array. Results are placed in A-array.  *
C                                                              *
C     Calculate:    PT    : total points.                      *
C                   VvPR  : Volume fraction parenchyma.        *
C                   VvNP  : Volume fraction non-parenchyma.    *
C                                                              *
C         AN    : Animal number (integer).                     *
C         BN    : Block number (integer).                      *
C         VOL   : Lung volume in cc (real).                    *
C         A     : Array for blocks from one animal.            *
C         B     : Array holds tally data and totals for blocks.*
C         NF    : Number of fields for block (integer).        *
C         LT    : Total line length in microns (real).         *
C         FLAG  : Flag set when data processed (logical*1)     *
C         NBLKS : number of cumulative blocks for this animal  *
C                                                              *
C******************************************************************

      LOGICAL*1  FLAG
      INTEGER    AN,BN,NF,NL,NBLKS
      REAL       D,VOL,LT,LTO,VPR
      REAL*8     L(30)

      COMMON     A(20,32),B(61,30),G(18,33)

C  define variable names for output file:
      L(1)='PT     '
      L(2)='VvPR   '
      L(3)='VvNP   '

      NL=3                  ! number of calculations for final output.
      FLAG=.TRUE.           ! let CBLOCK.FOR know data has been processed.

C  increment block counter:
      NBLKS=NBLKS+1

C  total block data:
      DO 50 J=1,2
        DO 60 I=1,NF
          B(NF+1,J)=B(NF+1,J)+B(I,J)
60      CONTINUE
50    CONTINUE

C  calculate PT and volume fractions and place results in A-array:
      A(NBLKS,1)=B(NF+1,1)+B(NF+1,2)        ! PT
      IF( A(NBLKS,1).EQ.0.0 ) GOTO 200      ! check for div by 0.
      A(NBLKS,2)=B(NF+1,1)/A(NBLKS,1)       ! VvPR
      A(NBLKS,3)=B(NF+1,2)/A(NBLKS,1)       ! VvNP

      A(NBLKS,31)=BN
      A(NBLKS,32)=NF

      RETURN                ! normal exit.

200   STOP 'Divide by zero (PT=0) in LMO.FOR'

      END
```

A Guide to Practical Stereology

```
              SUBROUTINE LM2(AN,BN,VOL,NF,LTO,D,L,NL,FLAG,NBLKS)
C******************************************************************
C      LM2.FOR                                    EFR 24-JUN-80    *
C                                                 NKT 26-JUL-81    *
C                                                                  *
C      Perform data calculations for LM2 stereology data on the    *
C   block of data in B-array.                                      *
C                                                                  *
C      Column definitions for B-array:                             *
C          1 : Pa, alveolar lumen                                  *
C          2 : Pad, alveolar duct lumen                            *
C          3 : Pt, interalveolar septal tissue                     *
C          4 : Pvlu, venule lumen                                  *
C          5 : Pvw, venule wall                                    *
C          6 : Palu, arteriolar lumen                              *
C          7 : Paw, arteriolar wall                                *
C          8 : Pbrl, bronchiolar lumen                             *
C          9 : Pbw, bronchiolar wall                               *
C         10: Pbl, bronchial lumen                                 *
C         11: Pbrw, bronchial wall                                 *
C         12: Prbl, resp. bronchiolar lumen                        *
C         13: Prbw, resp. bronchiolar wall                         *
C         14: Pfc, free cells                                      *
C         15: It, septal tissue intersections                      *
C         16: Iaw, arteriolar wall intersections                   *
C         17: Ivw, venular wall intersections                      *
C         18: Ibw, bronchiolar wall intersections                  *
C         19: Irbw, resp. bronchiolar wall intersections           *
C                                                                  *
C      Column definitions for Array:                               *
C          1 : PT, total points counted                            *
C          2 : Pit, septal tissue intersections                    *
C          3 : Vvt, volume fraction septal tissue                  *
C          4 : Vvtt, v.f. total tissue                             *
C          5 : Vvta, v.f. total airspace                           *
C          6 : Vvprta, v.f. total parenchymal airspace             *
C          7 : Vvnpta,v.f. total nonparenchymal airspace           *
C          8 : Vvnpt, v.f. nonparenchyma                           *
C          9 : Vvtvl, v.f. vessel lumen                            *
C         10: Vvpr, v.f. parenchyma                                *
C         11: Vvtvw, v.f. vessel wall                              *
C         12: MLI, mean linear intercept (mic)                     *
C         13: R, mean alveolar radius                              *
C         14: T, mean alveolar septal thickness                    *
C         15: Svpr, total parenchymal airspace                     *
C         16: Spr, parenchymal surface area                        *
C                                                                  *
C          AN    : Animal number (integer).                        *
C          BN    : Block number (integer).                         *
C          VOL   : Lung volume in cc (real).                       *
C          A     : Array for blocks for one animal.                *
C          B     : Array holds tally data and totals for blocks.   *
C          NF    : Number of fields for block (integer).           *
C          LT    : Total line length in microns (real).            *
C          FLAG  : Flag set when data processed (logical*1)        *
C          NBLKS : Number of blocks for this animal.               *
C                                                                  *
C******************************************************************
              LOGICAL*1    FLAG
              INTEGER      AN,BN,NF,NL,NBLKS
```

Programs for Estimating Stereological Values of Mammalian Lung and Heart

```
          REAL           D,VOL,LT,H,PC,LTO
          REAL*8         L(30)

          COMMON         A(20,32),B(61,30),G(18,33)

          H=1.0
          PC=1.0
C     define variable names for GRPRES subroutine:
          L(1)='PT    '
          L(2)='Pit   '
          L(3)='Vvt   '
          L(4)='Vvtt  '
          L(5)='Vvta  '
          L(6)='Vvprta'
          L(7)='Vvnpta'
          L(8)='Vvnpt '
          L(9)='Vvtvl '
          L(10)='Vvpr  '
          L(11)='Vvtvw '
          L(12)='MLI   '
          L(13)='R  (mic)'
          L(14)='T  (mic)'
          L(15)='Svpr  '
          L(16)='Spr   '

          NL=16
          FLAG=.TRUE.    ! let CBLOCK.FOR know data has been processed.

C     increment block counter:
          NBLKS=NBLKS+1

C     total block data:
          DO 50 J=1,30
             DO 60 I=1,NF
                B(NF+1,J)=B(NF+1,J)+B(I,J)
60           CONTINUE
50        CONTINUE

C     calculate PT and volume fractions and place results in A-array:
          DO 100 J=1,14
             A(NBLKS,1)=A(NBLKS,1)+B(NF+1,J)                  ! PT
100       CONTINUE
          IF( A(NBLKS,1).EQ.0.0 ) GOTO 200                    ! check for div
          LT=LTO*D*A(NBLKS,1)
          A(NBLKS,2)=B(NF+1,15)                               ! Pit
          A(NBLKS,3)=B(NF+1,3)*H/A(NBLKS,1)                   ! Vvt
          A(NBLKS,4)=(B(NF+1,3)*H+B(NF+1,5)+B(NF+1,7)+B(NF+1,9)
         *           +B(NF+1,11)+B(NF+1,13)+B(NF+1,14))/A(NBLKS,1) ! Vvtt
          A(NBLKS,5)=((B(NF+1,1)+B(NF+1,2)+B(NF+1,12))/
         *           H+B(NF+1,8)+B(NF+1,10))/A(NBLKS,1)       ! Vvta
          A(NBLKS,6)=(B(NF+1,1)+B(NF+1,2)+B(NF+1,12))
         *           /(H*A(NBLKS,1))                          ! Vvprta
          A(NBLKS,7)=(B(NF+1,8)+B(NF+1,10))/A(NBLKS,1)        ! Vvnpta
          A(NBLKS,8)=(B(NF+1,5)+B(NF+1,7)+B(NF+1,9)+B(NF+1,11))
         *           /A(NBLKS,1)                              ! Vvnpt
          A(NBLKS,9)=(B(NF+1,4)+B(NF+1,6))/A(NBLKS,1)         ! Vvtvl
          A(NBLKS,10)=1.0-A(NBLKS,7)-A(NBLKS,8)-A(NBLKS,9)    ! Vvpr
          A(NBLKS,11)=(B(NF+1,5)+B(NF+1,7))/A(NBLKS,1)        ! Vvtvw
          IF ( A(NBLKS,2).EQ.0.0) GO TO 110
          A(NBLKS,12)=LT*PC/A(NBLKS,2)                        ! MLI
          A(NBLKS,13)=0.75*A(NBLKS,12)                        ! R
          A(NBLKS,14)=(D*B(NF+1,3)*PC)/(4.0*A(NBLKS,2))       ! T
```

```
              A(NBLKS,15)=4.0*A(NBLKS,2)/(LT*PC)            ! Svpr
              A(NBLKS,16)=VOL*A(NBLKS,15)*A(NBLKS,10)       ! Spr
110           A(NBLKS,31)=BN
              A(NBLKS,32)=NF

              RETURN            ! normal exit.

200           STOP 'Divide by zero (PT=0) in LM2.FOR'

              END
```

Programs for Estimating Stereological Values of Mammalian Lung and Heart 283

```
          SUBROUTINE EM1(AN,BN,VOL,NF,LTO,D,L,NL,FLAG,NBLKS,VPR)

C*******************************************************************
C     EM1.FOR                                    EFR 24-JUN-80    *
C                                                NKT 23-JUL-81    *
C                                                                 *
C     Perform data calculations for EM1 stereology data on the    *
C     totaled block of data in B-array.                           *
C                                                                 *
C          Column definitions for B-array:                        *
C             1 : Pa, alveolar lumen                              *
C             2 : Pep1, epithelial type 1 cell                    *
C             3 : Pep2, epithelial type 2 cell                    *
C             4 : Pcen, capillary endothelium                     *
C             5 : Pcl, capillary lumen                            *
C             6 : Pbi, barrier interstitium                       *
C             7 : Psi, septal interstitium                        *
C             8 : Ps, septal cells                                *
C             9 : Pam, alveolar macrophage                        *
C            10: Iep1, type 1 cell intersections                  *
C            11: Iep2, type 2 cell intersections                  *
C            12: Ienw, endothelial cell intersections             *
C                                                                 *
C          Column definitions for A-array:                        *
C             1 : PT/block, total points counted                  *
C             2 : Vva, volume fraction alveolar air space         *
C             3 : Vvep1, v.f. type 1 epithelium                   *
C             4 : Vvep2, v.f. type 2 epithelium                   *
C             5 : Vvcen, v.f. capillary endothelium               *
C             6 : Vvcl, v.f. capillary lumen                      *
C             7 : Vvbi, v.f. barrier interstitium                 *
C             8 : Vvsi, v.f. septal interstitium                  *
C             9 : Vvs, v.f. septal cells                          *
C            10: Vvam, v.f. alveolar macrophages                  *
C            11: Vvti, v.f. total interstitium                    *
C            12: Sva, surface to vol. ratio of alveolar epithelium*
C            13: Svcen, surface to vol. ratio cap endothel. wall  *
C            14: Tep1,ep2, arithmetic mean thickness epithelium   *
C            15: Tenw, a.m.t. capillary wall                      *
C            16: Tti, a.m.t. interstitium                         *
C            17: Ttb, a.m.t. total blood air-barrier              *
C            18: Sa, alveolar surface area of fixed lung          *
C            19: Sc, capillary surface area of fixed lung         *
C                                                                 *
C          AN    : Animal number (integer).                       *
C          BN    : Block number (integer).                        *
C          VOL   : Lung volume in cc (real).                      *
C          A     : Array for blocks from one animal.              *
C          B     : Array holds tally data and totals for blocks.  *
C          NF    : Number of fields for block (integer).          *
C          LT    : Total line length in microns (real).           *
C          FLAG  : Flag set when data processed (logical*1)       *
C          NBLKS : Number of cumulative blocks for this animal    *
C                                                                 *
C*******************************************************************

          LOGICAL*1  FLAG
          INTEGER    AN,BN,NF,FLIMIT,NL,NBLKS
          REAL       D,VOL,LT,LTO,H,PC,VPR
          REAL*8     L(30)

          COMMON     A(20,32),B(61,30),G(18,33)
```

```
              H=1.0                            ! these values can be changed to
              PC=1.0                           ! correct for processing and
C                                              ! Holmes effect

C     define variable names for GRPRES subroutine:
              L(1)='PT/block'
              L(2)='Vva      '
              L(3)='Vvep1    '
              L(4)='Vvep2    '
              L(5)='Vvcen    '
              L(6)='Vvcl     '
              L(7)='Vvbi     '
              L(8)='Vvsi     '
              L(9)='Vvs      '
              L(10)='Vvam    '
              L(11)='Vvti    '
              L(12)='Sva     '
              L(13)='Svcen   '
              L(14)='Tep1,ep2'
              L(15)='Tenw    '
              L(16)='Tti     '
              L(17)='Ttb     '
              L(18)='Sa      '
              L(19)='Sc      '

              NL=19
              FLAG=.TRUE.      ! let CBLOCK.FOR know data has been processed.

C     increment block counter:
              NBLKS=NBLKS+1

C     total block data:
              DO 50 J=1,30
                  DO 60 I=1,NF
                      B(NF+1,J)=B(NF+1,J)+B(I,J)
60                CONTINUE
50            CONTINUE


C     calculate PT and volume fractions and place results in A array:
              DO 100 J=1,9
                  A(NBLKS,1)=A(NBLKS,1)+B(NF+1,J)
100           CONTINUE
              IF( A(NBLKS,1).EQ.0.0 ) GOTO 200        ! check for div by 0.
              LT=LTO*D*A(NBLKS,1)                     ! LT
              A(NBLKS,2)=B(NF+1,1)/A(NBLKS,1)         ! Vva
              A(NBLKS,3)=B(NF+1,2)/A(NBLKS,1)         ! Vvep1
              A(NBLKS,4)=B(NF+1,3)/A(NBLKS,1)         ! Vvep2
              A(NBLKS,5)=B(NF+1,4)/A(NBLKS,1)         ! Vvcen
              A(NBLKS,6)=B(NF+1,5)/A(NBLKS,1)         ! Vvcl
              A(NBLKS,7)=B(NF+1,6)/A(NBLKS,1)         ! Vvbi
              A(NBLKS,8)=B(NF+1,7)/A(NBLKS,1)         ! Vvsi
              A(NBLKS,9)=B(NF+1,8)/A(NBLKS,1)         ! Vvs
              A(NBLKS,10)=B(NF+1,9)/A(NBLKS,1)        ! Vvam
              A(NBLKS,11)=(B(NF+1,6)+B(NF+1,7)+B(NF+1,8))
          *              /A(NBLKS,1)                  ! Vvti
              A(NBLKS,12)=(2.0*(B(NF+1,10)+B(NF+1,11)))
          *              /LT*PC                       ! Sva
              A(NBLKS,13)=(2.0*(B(NF+1,12)))/LT*PC    ! Svcen
              IF ((B(NF+1,10)+B(NF+1,11)).EQ.0.0) GO TO 150
              A(NBLKS,14)=(D*(B(NF+1,2)+B(NF+1,3))*PC)
          *              /(4.0*(B(NF+1,10)+B(NF+1,11)))  ! Tep1,ep2
```

```
150       IF (B(NF+1,12).EQ.0.0) GO TO 160
          A(NBLKS,15)=(D*B(NF+1,4)*PC
     *               /(4.0*B(NF+1,12))               ! Tenw
160       IF ((B(NF+1,10)+B(NF+1,11)+B(NF+1,12)).EQ.0.0)
     *           GO TO 170
          A(NBLKS,16)=(D*(B(NF+1,6)+B(NF+1,7)
     *               +B(NF+1,8)))/(2.0*(B(NF+1,10)
     *               +B(NF+1,11)+B(NF+1,12)))        ! Tti
          A(NBLKS,17)=(D*(B(NF+1,2)+B(NF+1,3)
     *               +B(NF+1,4)+B(NF+1,6)+B(NF+1,7)
     *               +B(NF+1,8))*PC)/(2.0*(B(NF+1,10)
     *               +B(NF+1,11)+B(NF+1,12)))        ! Ttb
170       A(NBLKS,18)=VOL*VPR*A(NBLKS,12)            ! Sa
          A(NBLKS,19)=VOL*VPR*A(NBLKS,13)            ! Sc

          A(NBLKS,31)=BN
          A(NBLKS,32)=NF

          RETURN            ! normal exit.

200       STOP 'Divide by zero (PT=0) in EM1.FOR'

          END
```

```
c                 this program is an interactive data entry program
c                      to accept capillary axial ratio and density data
c                                            written by Jim inderbitzen 1-13-81
      character a,fname * 14,buf * 60
      integer r1,r2,freq(7)
      write(6,300)
  300 format(/,15x,'axial ratio program - version 2.1 ')
      write(6,301)
  301 format(/,2x,'would you like instructions?(y/n)')
      read(5,600)a
      if (a .ne. 'y') goto 205
      open(3,file='capins',status='old',form='form',access='seq')
      rewind(3)
      do 800 j=1,23
         read(3,801)line
  801    format(a)
  800    write(6,801)buf
  205 write(6,304)
  304 format(/,2x,'enter filename for this region',/,5x,
     1        ' use this format : reg..an..',/,5x,
     2          '{example : res31an13   means animal # 13,',
     3           ' region 31}',/,6x,'-- maximum of 9 characters')
      read(5,610)fname
  610 format(a)
      open(2,file=fname)
      write(6,469)
  469 format(/,3x,'is this the beginning of a region?(y/n)')
      read(5,600)a
      if(a .eq. 'n') goto 288
      write(2,701)fname
  701 format(a)
      write(6,470)
  470 format(/,3x,'how many blocks for this region?')
      read*,nblock
      write(6,471)
  471 format(/,3x,'how many fields per block?')
      read*,nfield
      write(2,700)nblock,nfield
  700 format(2i3)
  288 write(6,410)
  410 format(/,3x,'enter test area for this region',
     1        ' (in sq. mm.)')
      read*,atest
      write(6,411)
  411 format(/,3x,'enter class width for this region',
     1        ' (in microns)')
      read*,width
c                         new field
  230 write(6,310)
  310 format(/,5x,'new field')
c             set all field variables to zero
  277 fnmeq = 0.0
      fsmeq = 0.0
      nmeq = 0
      fsmeq2 = 0.0
      fsumr = 0.0
      fsumr2 = 0.0
      ifn = 0
      do 120 i=1,7
  120    freq(i) = 0
      write(6,380)
  380 format(/,3x,'enter field name - examples: a1,c2,f1')
      read(5,381)field
  381 format(a2)
      write(6,413)
  413 format(/,3x,'enter total number of capillaries for this field')
```

```
            read*,numcap
            write(6,415)
     415 format(/,3x,'is test area for this field the same as'
           1           ' last test area entered? (y/n)')
            read(5,600)a
            if(a .eq. 'y') goto 240
            write(6,420)
     420 format(/,2x,'enter new test area (sq. mm.)')
            read*,atest
c                       enter axial ratios
     240 write(6,340)
     340 format(5x,'enter r1,r2,how many')
            read*,r1,r2,ndo
c                       end of data for this field?
            if (r1 .eq. 0) goto 254
            ratio = float(r1)/float(r2)
            ifn = ifn + ndo
c                       sum into applicable categories
            do 100 k=1,ndo
               if(r1 .ne. r2) goto 250
c                       ratios of 1/1
               freq(r1) = freq(r1) + 1
               nmeq = nmeq + 1
               fsmeq = fsmeq + r1
               fsmeq2 = fsmeq2 + r1 ** 2.0
c                       all ratios
     250    fsumr = fsumr + ratio
            fsumr2 = fsumr2 + ratio ** 2.0
     100 continue
            goto 240
     254 if(numcap .eq. ifn)goto 255
c                       mistake in entering data
            write(6,444)
     444 format(/,3x,'you screwed up, you good-for-nothing ',
           1        'work study employee',/,6x,'number of capillaries',
           2        ' entered at beginning of field does not agree ',
           3        /,6x,'with total number of ratios entered - - ',
           4        're-enter this field',/,9x,'and do it right this time')
c                       go back and re-enter field
            goto 277
c                       compute capillary profile density
     255 cpd = fsumr/atest
c                       write to file
            write(2,705)field
     705 format (a2)
            write(2,710)ifn,fsumr,fsumr2,cpd
     710 format(i3,3f12.5)
            write(2,715)nmeq,fsmeq,fsmeq2,width
     715 format(i3,3f12.5)
            write(2,720)(freq(j),j=1,7)
     720 format(7i3)
     260 write(6,350)
     350 format(/,15x,'another field for this block?(y/n)')
            read(5,600)a
     600 format(a1)
            if (a .eq. 'y') goto 230
c                       end of region
            close(2)
     270 write(6,400)
     400 format(///,4x,'another region for this animal?(y/n)')
            read(5,600)a
            if(a .eq. 'y') goto 205
            write(6,456)
     456 format(/,5x,'all done!')
            end
```

```
c             this is an interactive data entry program that computes
c                volumetric densities of various components of the heart.
c                the program creates a data file to be used in the future
c                for statistical analysis.
c                                    written by jim inderbitzen 6-81
      dimension d(47),mmess(6),labels(34),loops(4,2)
      logical check
      character fname*14,a,a1,a2,grid*4,labels*6,blk,bmess(6),blname(8)
      write(6,400)
  400 format(//,10x,'volumetric density data entry',//)
   10 write(6,401)
  401 format(5x,'enter file name - example: vol10an13')
      read(5,300)fname
  300 format(a)
      open(2,file=fname)
      write(6,405)
  405 format(/,3x,'is this a new file?(y/n)')
      read(5,305)a1
  305 format(a1)
   30 write(6,435)
  435 format(/,'enter grid type { a100,b100,c64,d64,l100,m168 }',
     1        '  for all levels')
      read(5,301)grid
  301 format(a4)
      check = .true.
      call gtype(grid,check,ak1,ak2)
      if (check .eq. .false.)goto 30
      if(a1 .eq. 'n')goto 20
c                         new file
      write(2,702)fname
  702 format(a)
      write(6,410)
  410 format(3x,'is this region complete -- a consistent # of fields',
     1        '  per block ? {y/n}')
      read(5,305)a
      write(6,430)
  430 format(/,'enter standard number of blocks & number of fields ')
      read*,numblk,numfld
      write(6,431)
  431 format('enter blocks that data will be entered for -- '
     #        '  example: abcdfgh')
      read(5,307)(blname(j),j=1,numblk)
  307 format(8a1)
      numsk = 0
      do 102 k=1,6
        bmess(k) = ' '
  102   mmess(k) = 0
      if(a .eq. 'y')goto 25
c                              incomplete blocks
      write(2,705)
  705 format('incomplete')
      write(6,415)
  415 format('how many blocks have missing fields?')
      read*,numsk
      write(6,417)
  417 format('for these incomplete blocks ...')
      do 100 m=1,numsk
        write(6,420)
  420   format(2x,'enter block name and number of fields ',
     #          'in this block -- {example: a,4}')
  303   format(a1,1x,i1)
  100   read(5,303)bmess(m),mmess(m)
      nrec = (numblk - numsk) * numfld
      do 101 k=1,numsk
  101   nrec = nrec + mmess(k)
```

Programs for Estimating Stereological Values of Mammalian Lung and Heart 289

```
            goto 33
c                                 complete
   25 write(2,720)
  720 format('complete')
      nrec = numblk * numfld
   33 write(2,710)nrec,numblk,numfld,(blname(j),j=1,numblk),numsk
  710 format(3i3,3x,8a1,3x,i3)
      write(2,715)(bmess(k),mmess(k),k=1,numsk)
  715 format(6(a1,i2,2x))
      if(a1 .eq. 'n')goto 20
      write(2,725)grid,ak1,ak2
  725 format(a4,2f10.6)
c                       access data label file
   20 open(3,file='datalabeloops',status='old',
     *     access='sequential',form='formatted')
      rewind(3)
      read(3,600)(labels(k),k=1,34)
  600 format(a4)
      read(3,605)((loops(j,k),k=1,2),j=1,4)
  605 format(4(2i2,1x))
      close(3)
      write(6,436)
  436 format(3x,'would you like to enter data for levels 1-4 (y)',/,
     1      5x,'or just levels 1 & 2 (n)   ?')
      read(5,305)a2
      levels = 2
      nwrite = 47
      if(a2 .eq. 'y') levels = 4
      if(a1 .eq. 'n') goto 22
      write(2,726)levels,nwrite
  726 format('data for levels 1-',i1,' - ',i2,' variables')
c                       new block
   22 write(6,439)
  439 format('enter block name  (examples : a,d}')
      read(5,305)blk
      write(6,437)(labels(m),m=1,8)
  437 format('enter data for this block :',/,8(2x,a6))
      read*,(d(k),k=1,8)
c                       new field
   40 write(6,440)
  440 format('enter field number  (examples : 1,4}')
      read(5,310)fld
  310 format(i1)
      write(2,727)blk,fld
  727 format(a1,i1)
      write(6,442)blk,fld
  442 format(' enter data for field ',a1,i1,' ...')
      do 105 j=2,levels
        n1 = loops(j,1)
        n2 = loops(j,2)
        write(6,446)
  446   format()
        write(6,445)(labels(m),m=n1,n2)
  445   format(11(3x,a4))
  105   read*,(d(k),k=n1,n2)
c                       computation of variables
      call comput(d,ak1,ak2,levels)
c             write field to file
      write(2,730)(d(j),j=1,nwrite)
  730 format(1p,5e12.5)
   50 write(6,450)
  450 format('enter another field for this block?(y/n)')
      read(5,305)a
      if(a .eq. 'y') goto 40
      write(6,455)
  455 format('enter another block for this region?(y/n)')
```

```
      read(5,305)a
      if(a .eq. 'y') goto 22
      close(2)
      write(6,460)
  460 format('enter another region?(y/n)')
      read(5,305)a
      if(a .eq. 'y')goto 10
      write(6,465)
  465 format('all done!!')
      end
c           ........................
c           .   subroutine stype   .
c           ........................
      subroutine stype(grid,check,ak1,ak2)
      character grid*4
      logical check
      if(grid .eq. 'a100')goto 35
      if(grid .eq. 'b100')goto 35
      if(grid .eq. 'c64')goto 35
      if(grid .eq. 'd64')goto 35
      if(grid .eq. 'l100')goto 36
      if(grid .eq. 'm168')goto 37
      check = .false.
      write(6,437)
  437 format('illegal grid type!!!')
      return
   35 ak1 = 1.0
      ak2 = 1.0
      return
   36 ak1 = 1.570796
      ak2 = 1.0
      return
   37 ak1 = 0.5
      ak2 = 0.866025
      return
      end
c           ........................
c           .   subroutine comput  .
c           ........................
      subroutine comput(d,ak1,ak2,levels)
      dimension d(47)
c                       level 1
      d(35) = d(4) + d(5) + d(6) + d(7) + d(8)
      d(36) = 1.0 - d(35)
c                       level 2
      d(37) = d(10) + d(11) + d(12) + d(13)
      d(38) = d(37) - d(10)
      d(39) = d(37) * ak1 * d(9) * d(1) * d(2)
      d(40) = d(37) * ak2 * (d(9) * d(1) * d(2))**2.0
c                       if only entering data for levels 1 & 2,
c                           skip remaining computations
      if(levels .eq. 2) return
c                       level 3
      d(41) = d(16) + d(17) + d(18) + d(19) + d(20) + d(21) + d(22)
      d(42) = d(41) * ak1 * d(15) * d(1) * d(2)
      d(43) = d(41) * ak2 * (d(15) * d(1) * d(2))**2.0
      d(44) = 0.25 * d(15) * d(20) * d(1) * d(2) / d(23)
c                       level 4
      d(45) = d(25) + d(26) + d(27) + d(28) + d(29) +d(30) +d(31) +d(32)
      d(46) = d(45) * ak1 * d(24) * d(1) * d(2)
      d(47) = d(45) * ak2 * (d(24) * d(1) * d(2))**2.0
      return
      end
```

Bibliography

Abercrombie, M.: Estimation of nuclear population from microtome sections. Anat. Rec. *94:* 239–247 (1946).

Anversa, P.; Loud, A.V.; Giacomelli, F., and Wiener, J.: Absolute morphometric study of myocardial hypertrophy in experimental hypertension II. Ultrastructure of myocytes and interstitium. Lab. Invest. *38:* 597–609 (1978).

Bach, G.: Uber die Grössenverteilung von Kugelschnitten in durchsichtigen Schnitten endlicher Dicke. Z. wiss. Mikrosk. *64:* 265–270 (1959).

Bach, G.: Gründung einer internationalen Gesellschaft für Stereologie. Z. wiss. Mikrosk. *65:* 190–192 (1963a).

Bach, G.: Die Vernachlässigung der Schnittdicke histologischer Präparate als Ursache für Fehlauswertungen. *65:* 194–195 (1963b).

Bach, G.: Über die Bestimmung von charakteristischen Grössen einer Kugelverteilung aus der Verteilung der Schnittkreise. Z. wiss. Mikrosk. *65:* 285–291 (1963c).

Bach, G.: Bestimmung der Häufigkeitsverteilung der Radien kugelförmiger Partikel aus der Häufigkeiten ihrer Schnittkreise in zufälligen Schnitten der Dicke delta. Z. wiss. Mikrosk. *66:* 193–200 (1964).

Bach, G.: Über die Bestimmung von charakteristischen Grössen einer Kugelverteilung aus der Verteilung der Schnittkreise. Metrika *9:* 228–234 (1965).

Banchoff, T.F. and Strauss, C.M.: Real-time computer graphics analysis of figures in four-space; in Bresson, Hypergraphics, Visualizing Complex Relationships in Art, Science and Technology (Westview, Boulder 1978).

Bedrossian, C.W.M.; Anderson, A.E. Jr., and Foraker, A.G.: Comparison of methods for quantitating bronchial morphology. Thorax *26:* 406–408 (1971).

Bondi, H.: Cosmology. (Cambridge University, London 1960).

Born, G.: Die Plattenmodelliermethode. Archiv für mikroskopische Anatomie, *22:* 584–599 (1883).

Boyde, A.: Quantitative photogrammetric analysis and qualitative stereoscopic analysis of SEM images. J. Microsc. *98:* 452–471 (1973).

Boyde, A. and Ross, H.F.: Photogrammetry and the scanning electron microscope. Photogramm. Rec. *8:* 408–457 (1975).

Cayley, 1843, quoted by Sommerville, D.M.Y.: (original edition 1929). An introduction to the geometry of n dimensions; pp. 51–72 (Dover Pub., New York 1958).

Chalkley, H.W.: Method for quantitative morphological analysis of tissues. J. Natl. Cancer Inst. *4:* 47–53 (1943).

Cochran, W.G.: Sampling techniques (Wiley, New York 1963).

Cramer, H.: Mathematical methods of statistics (Princeton Univ., New Jersey 1946).

Crapo, J.D. and Greeley, D.A.: Estimation of the mean caliper diameter of cell nuclei. II. Various cell types in rat lung. J. Microsc. *114* (pt 1): 41–48 (1978).

Cruz-Orive, L.M.: Best linear unbiased estimators for stereology. Biometrics *36:* 595–605 (1980).

DeHoff, R.T.: The determination of the size distribution of ellipsoidal particles from measurements made on random plane sections. Trans. AIME *224:* 474–477 (1962).

DeHoff, R.T.: The relationship between mean surface curvature and the stereological counting measurements; in Elias, Stereology, pp. 95–105 (Springer, New York 1967).

DeHoff, R.T. and Bousquet, P.: Estimation of the size distribution of triaxial ellipsoidal particles from the distribution of linear intercepts. J. Microsc. *92:* 119–135 (1970).
DeHoff, R.T. and Rhines, F.N.: Determination of number of particles per unit volume from measurements made on random plane sections: The general cylinder and the ellipsoid. Trans. AIME *221:* 975–982 (1961).
DeHoff, R.T. and Rhines, F.N.: Quantitative microscopy (McGraw-Hill, New York 1968).
Delesse, A.: Procédé mécanique pour déterminer la composition des roches. Annls. Mines *13:* 379 (1848).
Dunn, R.F.; O'Leary, D.P., and Kumley, W.E.: Quantitative analysis of micrographs by computer graphics. J. Microsc. *105:* 205–213 (1975).
Ebbesson, S.O.E. and Tang, D.: A method for estimating the number of cells in histological sections. J.R. Microsc. Soc. *84:* 449–464 (1965).
Eisenberg, B.R.; Kuda, A.M., and Peter, J.B.: Stereological analysis of mammalian skeletal muscle. I. Soleus muscle of the adult guinea pig. J. Cell Biol. *60:* 732–754 (1974).
Elias, H.: Contributions to the geometry of sectioning. III. Spheres in masses. Z. wiss. Mikrosk. *62:* 32–40 (1954).
Elias, H.: Vierdimensionale Geometrie und ihre praktische Anwendung zur Erklärung kosmologischer Probleme. Experientia *12:* 362–364 (1956).
Elias, H.: De structura glomeruli renalis. Anat. Anz. *104:* 26–36 (1957).
Elias, H.: Structure and rotation of barred and spiral galaxies. Interpreted by methods of hyperstereology (i.e., extrapolation from n- to $(n + 1)$-dimensional space; in Elias, Stereology, pp. 149–159 (Springer-Verlag, New York 1967).
Elias, H.: Stereology of parallel, straight, circular cylinders. J. Microsc. *107:* 199–202 (1976).
Elias, H.: Ursprung der Stereologie und der internationalen Gesellschaft für Stereologie. Verh. Anat. Ges. *71:* 359–362 (1977).
Elias, H.: Thickness of a curved lamina. Microscope *28:* 67–73 (1980).
Elias, H. and Botz, E.: Simple devices for stereology and morphometry; in Underwood, de Wit, and Moore, Proc. 4th Int. Congr. Stereology, National Bureau of Standards, Special Pub. 431, pp. 431–434 (US Gov. Printing Off. 1976).
Elias, H.; Botz, E., and Hennig, A.: Einfache Vorrichtungen für Stereologie und Morphometrie. Verh. Anat. Ges. *70:* 1013–1017 (1976).
Elias, H. and Fong, B.B.: Nuclear fragmentation in colon carcinoma cells. Hum. Pathol. *9:* 679–684 (1978).
Elias, H. and Hennig, A.: Stereology of the human renal glomerulus; in Weibel and Elias Quantitative methods in morphology, pp. 130–166 (Springer, New York 1967).
Elias, H.; Hennig, A., and Elias, P.M.: Contributions to the geometry of sectioning. V. Some methods for the study of kidney structure. Z. wiss. Mikrosk. *65:* 70–82 (1961).
Elias, H.; Hennig, A., and Schwartz, D.E.: Stereology. Applications to biomedical research. Physiol. Rev. *51:* 158–200 (1971).
Elias, H.; Haug, H.; Lange, W.; Schlenska, G., and Schwartz, D.: Oberflächenmessungen der Grosshirnrinde von Saeugern mit besonderer Beruecksichtigung des Menschen, der Letacea, des Elephanten und der Marsupialia. Anat. Anz. *124:* 461–463 (1969).
Elias, H.; Hyde, D.M.; Mullens, R.S., and Lambert, F.C.: Colonic adenomas: Stereology and growth mechanisms. Dis. Colon Rectum *24:* 33–44 (1981).
Elias, H.; Pauly, J.E., and Burns, E.R.: Histology and human microanatomy; 4th ed. (Wiley, New York 1978).
Elias, H. and Schwartz, D.: Surface areas of the cerebral cortex of mammals determined by stereological methods. Science *166:* 111–113 (1969).
Elias, H. and Schwartz, D.: Cerebrocortical surface areas, volumes, lengths of gyri and their interdependence in mammals, including man. Z. Säugetierk. *36:* 147–163 (1971).

Bibliography

Elias, H. and Spanier, E.H.: Structure of the collagenous tissue in the cirrhotic liver, a contribution to the geometry of sectioning. Z. wiss. Mikrosk. *61:* 213–221 (1953).

Federle, M.P. and Moss, A.A.: Computed tomography of the spleen; in Margulis and Burhenne, Alimentary tract roentgenology (Mosby, St. Louis (in press) 1981).

Fisher, C.: The new Quantimet 720. Microscope *19:* 1–20 (1971).

Forsyth, A.R.: Geometry of four dimensions (Cambridge Univ., Cambridge 1930).

Froberg, C.E. Introduction to numerical analysis (Addison-Wesley, Reading, Massachusetts 1965).

Gaunt, W.A. and Gaunt, P.N.: Three dimensional reconstruction in biology (University Park, Baltimore 1978).

Gazzaniga, M.S.: One brain—two minds. Am. Sci. *60:* 311–317 (1972).

Gerdes, A.M.; Callas, G., and Kasten, F.H.: Differences in regional capillary distribution and myocyte sizes in normal and hypertrophic rat hearts. J. Anat. *156:* 523–531 (1979).

Glagolev, A.A.: Quantitative analysis with the microscope by the point method. Eng. Min. J. *135:* 399–402 (1934).

Greeley, D.; Crapo, J.D., and Vollmer, R.T.: Estimation of the mean caliper diameter of cell nuclei. I. Serial section reconstruction method and endothelial nuclei from human lung. J. Microsc. *114* (pt. 1): 31–39 (1978).

Gundersen, H.J.G.; Jensen, T.B., and Østerby, R.: Distribution of membrane thickness determined by lineal analysis. J. Microsc. *113:* 27–43 (1977).

Haug, H.: Probleme und Methoden der Strukturzählung im Schnittpräparat; in Weibel and Elias, Quantitative methods in morphology, pp. 58–78 (Springer, New York 1967).

Hegre, E.S. Brashear, A.D.: Block-surface staining and cinematography. Stain Techn. *21:* 161–164 (1946).

Hennig, A.: Länge eines räumlichen Linienzuges. Z. wiss. Mikrosk. *65:* 193–194 (1963a).

Hennig, A: Length of a three-dimensional linear tract; in Proc. 1st Intl. Congr. Stereology pp. 44/1–44/8 (Vienna 1963b).

Hennig, A. and Elias, H.: Contributions to the geometry of sectioning. VI. Theoretical and experimental investigations on sections of rotatory ellipsoids. Z. wiss. Mikrosk. *65:* 133–145 (1963a).

Hennig, A. and Elias, H.: Sections through triaxial ellipsoids; in Proc. 1st Int. Congr. Stereology, pp. 43/1–43/12 (Vienna, 1963b).

Hennig, A. and Elias, H.: Untersuchung von Kalottenanteilen aus Schnittbildern und Tangierproblem von Kernkörperchen. Mikroskopie *21:* 32–36 (1966).

Hennig, A. and Elias, H.: A rapid method for the visual determination of size distribution of spheres from the size distribution of their sections. J. Microsc. *93:* 101–107 (1970).

Herrmann, H.J.; Mühlig, P.; Kuhne, C., and Läuter, J.: Automated image analysis for measurements of morphological reactions of blood vessels of the microvascular system. Exp. Pathol. *17:* 215–227 (1979).

Heymsfield, S.B.; Fulenwider, T.; Nordlinger, B.; Barlow, R.; Sones, P., and Kutner, M.: Accurate measurement of liver, kidney, and spleen volume and mass by computerized axial tomography. Ann. Intern. Med. *90:* 185–187 (1979).

Hilliard, J.E.: Specification and measurement of microstructural anisotropy. Trans. AIME *224:* 1201–1211 (1962).

Hilliard J.E. and Chan, J.W.: An evaluation of procedures in quantitative metallography for volume fraction analysis. Trans. AIME *221:* 344–356 (1961).

Hoffmann, H.P. and Avers, C.J.: Mitochondrion of yeast. Science *181:* 749–750 (1973).

Holmes, A.: Petrographic methods and calculations (Thomas Murby, London 1921).

Hubble, E.: Effects of red shifts on the distribution of nebulae. Astrophys. J. *84:* 517–554 (1936).

Hyde, D.; Orthoefer, J.; Dungworth, D.; Tyler, W.; Carter, R., and Lum, H.: Morphometric and morphologic evaluation of pulmonary lesions in beagle dogs chronically exposed to high ambient levels of air pollutants. Lab. Invest. *38:* 455–469 (1978).

Hyde, D.M.; Robinson, N.E.; Gillespie, J.R., and Tyler, W.S.: Morphometry of the distal air spaces in lungs of aging dogs. J. Appl. Physiol. *43:* 86–91 (1977).

Kastschenko, N.: Methode zur genaueren Rekonstruktion kleinerer makroskopischer Gegenstände. Arch. Anat. 388–394 (1886).

Kastschenko, N.: Ueber das Beschneiden mikroskopischer Objekte. Z. wiss. Mikrosk. *5:* 173–181 (1888).

Kaye, G.I.; Lane, N.; Wheeler, H.O., and Witlock, R.T.: Fluid transport in the rabbit gall bladder: A combined physiological and microscopic study. Anat. Rec. *151:* 369 (1965).

Keller, H.J. and Burri, P.H.: Automatic pattern analysis of growing rat lung. J. Microsc. *121:* 119–130 (1981).

Keller, H.J.; Friedli, H.P.; Gehr, P.; Bachofen, M., and Weibel, E.R.: The effects of optical resolution on the estimation of stereological parameters; in Underwood, DeWit, and Moore, Proc. 4th Int. Congr. Stereology, National Bureau of Standards, Special Pub. 431, pp. 409–410 (US Gov. Printing Off., Washington, D.C. 1976).

Koop, J.C.: Systematic sampling of two-dimensional surfaces and related problems, RTI Project No. 250U-1123 (Statistical Sciences Group, Research Triangle Institute, 1976).

Kovalevsky, J.: La galaxie. Sciences, rev. franc. *46:* 28–43 (1966).

Lackritz, J.R. and Schaefer, R.L.: Asymptotic properties of estimators derived from line transect sampling. Technical Report, Dept. of Statistics, Univ. Florida (1981).

Langston, C. and Thurlbeck, W.M.: The use of simple image analysers in lung morphometry. J. Microsc. *114* (1): 89–100 (1978).

Lenz, F.: Zur Grössenverteilung von Kugelschnitten. Z. wiss. Mikrosk. *63:* 50–56 (1956).

Loud, A.V.: Quantitative estimation of loss of membrane images resulting from oblique sectioning of biological membranes; in Areeneaux Proc. 25th EMSA, pp. 144–145 (Claitor's Book Store, Baton Rouge, Louisiana 1967).

Loud, A.V.; Anversa, P.; Giacomelli, F., and Wiener, J.: Absolute morphometric study of myocardial hypertrophy in experimental hypertension. I. Determination of myocyte size. Lab. Invest. *38:* 586–596 (1978).

Macagno, E.R.; Levinthal, C., and Sobel, I.: Three dimensional computer reconstruction of neurons and neuronal assemblies. Ann. Rev. Biophys. Bioeng. *8:* 323–351 (1979).

Mack, C.: On clumps formed when convex laminae or bodies are placed at random in two or three dimensions. Proc. Camb. Phil. Soc. *52:* 246–250 (1956).

Mathieu, O.; Cruz-Orive, L.M.; Hoppeler, H., and Weibel, E.R.: Measuring error and sampling variation in stereology: comparison of the efficiency of various methods for planar image analysis. J. Microsc. *121:* 75–88 (1981).

Mayall, N.U.: Comparison of rotational motions observed in the spirals M31 and M33 and in the galaxy. Publ. Obs. Univ. Michigan *10:* 19–24 (1951).

Mayall, N.U. and Aller, L.H.: The rotation of the spiral nebula Messier 33. Astrophys. J. *95:* 5–23 (1942).

Merz, W.A.: Die Streckenmessung an Gerichteten Strukturen im Mikroskop und ihre Anwendung zur Bestimmung von Oberflächen-Volumen-Relationen im Knochengewebe. Mikroskopie *22:* 132–142 (1967).

Miles, R.E. and Davy, P.: Precise and general conditions for the validity of a comprehensive set of stereological fundamental formulae. J. Microsc. *107:* 211–226 (1976).

Miles, R.E. and Davy, P.: On the choice of quadrants in stereology. J. Microsc. *110:* 27–44 (1977).

Mobley, B.A. and Page, E.: The surface area of sheep cardiac Purkinje fibers. J. Physiol. (Lond) *220:* 547–563 (1972).

Morton, R.R.A. and McCarthy, C.: The Omnicon^{tm} pattern analysis system. Microscope *23:* 239–260 (1975).

Moss, A.; Friedman, M.A., and Brito, A.C.: Determination of liver, kidney and spleen volumes by computed tomography: an experimental study in dogs. J. Comput. Assist. Tomogr. *5:* 12–14 (1981).

Nicholson, W.L.: Estimation of linear properties of particle size distributions. Biometrika *57:* 273–297 (1970).

Nicholson, W.L.: Application of statistical methods in quantitative microscopy. J. Microsc. *113:* 223–239 (1978).

Ohno, S.: Stereological application of thick sections to determination of size distribution of spherical cell organelles by high-voltage electron microscopy. J. Electron Microsc. *29:* 98–105 (1980a).

Ohno, S.: Morphometry for determination of size distribution of peroxisomes in thick sections by high-voltage electron microscopy: I. Studies on section thickness. J. Electron Microsc. *29:* 230–235 (1980b).

Ohno, S.; Yamabayashi, S.; Fujii, Y.; Kawahara, I., and Nagata, T.: Peroxisomal changes of rat livers during DEHP administration and after withdrawal. Acta Histochem. Cytochem. *12:* 611 (1979).

Oldendorf, W.H.: The quest for an image of brain: Computerized tomography in the perspective of past and future imaging methods (Raven, New York 1980).

Oliver, J.: Architecture of the kidney in chronic Bright's disease (Hoeber, New York 1939).

Oliver, J.: Nephrons and kidneys, a quantitative study of developmental and evolutionary mammalian architectonics (Hoeber, New York 1968).

Pannese, E.; Bianchi, R.; Calligaris, B.; Ventura, R., and Weibel, E.R.: Quantitative relationships between nerve and satellite cells in spinal ganglia. An electron microscopical study. I. Mammals. Brain Res. *46:* 215–234 (1972).

Pannesse, E.; Ventura, R., and Bianchi, R.: Quantitative relationships between nerve and satellite cells in spinal ganglia: An electron microscopical study. II. Reptiles. J. comp. Neurol. *160:* 463–476 (1975).

Peter, K.: Demonstration des Born-Peterschen Verfahrens zur Herstellung von Richtebenenen und Richtlinien. Anat. Anz. *16*, Ergh: 134–136 (1899).

Peter, K.: Die Methoden der Rekonstruktion. (Fischer, Jena 1906).

Peter, K.: Rekonstruktion in Schraegansicht. Z. wiss. Mikrosk. *39:* 138–148 (1922).

Phalen, R.F.; Yeh, H.C.; Schum, G.M., and Raabe, O.G.; Application of an idealized model to morphometry of the mammalian tracheobronchial tree. Anat. Rec. *190:* 167–176 (1978).

Pierce, R.J.; Brown, D.J.; Holmes, M.; Cumming, G., and Denison, D.M.: Estimation of lung volumes from chest radiographs using shape information. Thorax *34:* 726–734 (1979).

Plopper, C.G.; Halsebo, J.E.; Sonstegard, K.S., and Nettesheim, P.K.: Distribution of nonciliated bronchiolar epithelial (Clara) cells in intrapulmonary airways of rabbit lung. Am. Rev. Resp. Dis. *123:* 224 (1981).

Postlethwait, S.N.: Cinematography of serial microscope sections; in Proc. 1st Congr. Stereology, pp. 37a/1–37a/4 (Vienna 1963).

Raetz, H.U.; Gnaegi, H.R., and Weibel, E.R.: An on-line computer system for point counting stereology. J. Microsc. *101:* 267–282 (1974).

Rakusan, K.; Moraven, J., and Hatt, P.Y.: Regional capillary supply in the normal and hypertrophied rat heart. Microvasc. Res. *20:* 319–326 (1980).
Rosiwal, A.: Uber Geometrische Gesteinsanalysen; ein einfacher Weg zur Ziffermässigen Feststellung des Quantitätsverhältnisses der Mineralbestandtheile Gemengter Gesteine. (Verh. K.K. Geol. Reichsanst. Vienna, *6:* 143 (1898).
Saltykov, S.A.: Stereometric metallography, 2nd ed., pp. 446 (Metallurgizdat, Moscow, in Russian, 1958).
Scheaffer, R.L.: Sampling mixtures of multi-sized particles: An application of renewal theory. Technometrics *11:* 285–298 (1969).
Scheaffer, R.L.: An approximate variance for line intersection counts. J. Microsc. *103:* 343–349 (1975).
Schwartz, D.E. and Elias, H.: An optico-electrical particle size classifier. J. Microsc. *91:* 57–59 (1970).
Scudder, H.J.: Introduction to computer aided tomography. Proc. IEEE *66:* 628–637 (1978).
Sherwin, R.P.; Margolick, J.B., and Azen, S.P.: Hypertrophy of alveolar wall cells secondary to an air pollutant: A semi-automated quantitation. Arch. Environ. Health *26:* 297–300 (1973).
Sitte, H.: Morphometrische Untersuchungen an Zellen; in Weibel and Elias, Quantitative Methods in Morphology, pp. 167–198 (Springer-Verlag, New York 1967).
Small, J.V.: Measurement of Section Thickness; in Bocciarelli, Proceedings 4th European Congress on Electron Microscopy, vol. 1, pp. 609–610 (Tipographia Poliglotta Vaticana, Rome 1968).
Smith, C.S. and Guttman, L.: Measurement of internal boundaries in three-dimensional structures by random sectioning. Trans. AIME *197:* 81–111 (1953).
Snedecor, G.W. and Cochran, W.G.: Statistical Methods (Iowa State University, Ames 1967).
Sperber, I.: Studies on the mammalian kidney (Almquist, Uppsala 1944).
Stinson, S.F.,; Lilgan, J.; Reese, D.H.; Friedman, R.D., and Sporn, M.B.: Quantitation with an automated image analyzer of nuclear cytoplasmic changes induced by hydrocortisone in bladder epithelium. Cancer Res. *37:* 1428–1431 (1977).
Stinson, S.F. and Sporn, M.B.: Use of an automated image analyzer to quantitate cellular hyperplasia in urinary bladder epithelium. J. Microsc. *109:* 329–335 (1977).
Thurlbeck, W.M.: The internal surface area of nonemphysematous lungs. Am. Rev. Resp. Dis. *95:* 765–773 (1967).
Thurlbeck, W.M.: Post-mortem lung volumes. Thorax *34:* 735–739 (1979).
Underwood, E.E.: Particle Size Distribution; in De Hoff and Rhines, Quantitative Microscopy, pp. 149–200 (McGraw-Hill, New York 1968).
Underwood, E.E.: Quantitative stereology (Addison-Wesley, Reading, Massachusetts 1970).
Weibel, E.R.: Morphometry of the human lung (Springer, Berlin 1963).
Weibel, E.R.: Stereological principles for morphmetry in electron microscopic cytology. Int. Rev. Cytol. *26:* 235–302 (1969).
Weibel, E.R.: Morphometric estimation of pulmonary diffusion capacity. I. Model and method. Respir. Physiol. *11:* 54–75 (1970).
Weibel, E.R.: Stereological Methods; Vol. 1, Practical Methods for Biological Morphometry (Academic Press, New York 1979).
Weibel, E.R.: Stereological Methods; vol. 2, Theoretical Foundations. (Academic Press, New York 1980).
Weibel, E.R. and Elias, H.: Introduction to stereology and morphometry; in Weibel and Elias, Quantitative methods in morphology, pp. 3–19 (Springer, New York 1967).

Weibel, E.R. and Gomez, D.M.: A principle for counting tissue structures on random sections. J. Appl. Physiol. *17:* 343–348 (1962).

Weibel, E.R. and Knight, B.W.: A morphometric study on the thickness of the pulmonary air-blood barrier. J. Cell Biol. *21:* 367–384 (1964).

West, J.B. and Matthews, F.L.: Stress, strains and surface pressures in the lung caused by its weight. J. Appl. Physiol. *32:* 332–345 (1972).

Wicksell, S.D.: The corpuscle problem. I. A mathematical study of a biometric problem. Biometrika *17:* 84–99 (1925).

Woody, D.; Woody, E., and Crapo, J.D.: Determination of the mean caliper diameter of lung nuclei by a method which is independent of shape assumptions. J. Microsc. *118:* 421–427 (1980).

Subject Index

Airway identification technique 154
Angles
  amplitude 13
  complementary 10
  inclination 46
Anisodiametric particles
  definition 8
  histologic example,
    neurocranium 8
  mean caliper diameter 74
Anisotropy
  abolishing 120
  histologic example, skeletal
    muscle 116
  sampling methods 119
Archimedes principle, volume
  determination 28
Area
  absolute, solid 24
  alveolar surface 157
  basement membrane per
    glomerular volume 39
  curve 24
  filtration, symbol 39
  fraction, symbol 24
  measurements, equivalence 32
  surface
    estimate 162
    per unit volume
      isotropic body 37, 40
      symbol 24, 37
Areal sizing 130
Axial ratio
  calculation from traced
    coordinates 132, 133
  coronary capillaries,
    measurements, 171
  definition 23
  distribution of, cylinders 49, 52
Axis, intersection counting 116

Bias, avoiding 26, 27
"Big Bang" Theory 189, 194
Blood analysis by stereology 198

Caliper diameter, mean,
  prolate rotary 24, 77
Cardiac tissue
  capillary endothelium 172
  capillary surface density 168
  coronary arterial walls,
    arithmetic mean thickness 167
  capillaries
    axial ratio measurements 171
    mean diameter 172
    minimum diffusion
      distance 172
    profile density 172
    transmural orientation,
      potential 170
  microvasculature, method for
    estimating values 163, 168
  fixation and processing 163, 167
  Merz grid 168
  myocardial constituents
    hypertrophy 168
    symbols 173
  stereological treatment 163
  structural organization 163-165
  tissue selection 163, 166
  volumetric densities 167, 168
Cerebral cortex 119

Cerebrocortical surface area,
  absolute value, determination 40
Chord-sizing program 129
Circle, definition 3, 7
Circular cylinder 6
Cisternography, acoustic
  neuroma 133-135
Colonic adenoma
  conversion of colonic
    mucosa 177, 178
  definition 174
  epithelial area/mm^3 183
  epithelial surface area
    percent volume 183
  growth, definition 177
  nucleus, fragmentation 187, 188
  pedunculated polyps 185
  potential usefulness of
    stereology 174
  sessile polyps 185, 186
  table-shaped polyp 187
  volume, estimation 181, 182
Colonic mucosa
  number of epithelial cells,
    determination 180
  surface area, epithelial 178-180
  thickness 178-180
Columnar epithelium 7
Confidence intervals 100
Cosecant 10-12
Cosine,
  curve 12
  definition 10, 11
Cotangent, definition 10-12
Curvature
  definition 14
  degree of 14
  Gaussian 15
  mean surface, 15, 24
Cylinder
  axial ratios, distribution 49, 52
  branching, identification 53
  circular profiles, slices 48
  circular ratios in sections 50
  definition 6

directrix 6
elliptical
  axial ratio distribution 52
  description 52
  length and caliper diameter 79
right circular
  peripheral arterial valves 6
  truncated 79

DeHoff-Rhines-Underwood
  stereological symbols 23
Delesse principle
  equivalence of area
    measurements 32
  total volume of specimen 29
Density
  gradients 119
  numerical, concept 57
Diameter
  caliper 24, 77
  definitions 8
  symbols 24
Dimensional reduction
  laws of 20
  principles 20
  (see also section, slice, slab,
  circle and other shapes)
Distribution
  center 95
  estimator 95
  Q 189
  spread or variation 95

Electron microscopy
  Holmes effect 124
  thin features 124
  unbiased 26
Ellipse
  axial quotient 5
  description 4
  directrix of cylinder 6
  drawing 4
  focal points 4, 5
  oblique sections as 7

Subject Index

Ellipsoids
  oblate
    axial ratios, distribution 51
    description 50
    histologic examples 74, 75
    mean caliper diameter 78
    ratio of minor to major axis
      of cuts 75
  prolate
    axial ratio distribution 50, 51
    description 50
    histological examples 74, 75
    mean caliper diameter of
      selected solids 77
    ratio of minor to major axis
      of cuts 75
  rotary 6, 7
    axial ratio distribution 51
    caliper diameter
      method 75
      selected solids 77, 78
    description 50
    generation 7
    histologic examples 74, 75
  shape factor 76, 77
  triaxial
    caliper diameter, mean 78
    description 51
    endothelial nuclei 8
    histologic examples 75
    pulmonary type I epithelial
      cell nuclei 8
    shape determination 51, 52
    size distribution, method 78, 79
  types 6

Epididymidis
  length of duct, determination 42
  profile count of tubules 41
Erythrocytes
  number of cells/mm^3 of blood
  surface area per unit volume of
    blood 197

Fibroblast, nuclei, prolate rotatory
  ellipsoidal 7

Finite objects 120
Fixation
  processing, correction factors 156
  shrinkage, calculation 27
Folding, index of, symbol 24
Four dimensional
  model 191-194
  theory 195

Geometric concepts in stereology 7, 190
Gladolev, intercept measurements 32
Glomerulus
  basement membrane
    area/glomerular volume 39
    elasticity 200
    thickness 198
  filtration area 39, 199
Glycogen granules, electron opaque 124
Gray/white volume ratio 34
Gyri
  cerebral cortex, total length
    determination 44, 45
  orientation 45

Hematocrit, stereological approach 197
Histogram, composite 74
Histology, shapes encountered 9
Holmes effect 121-123, 147
  high power electron microscopy 124
  negative 190
  section thickness and 124, 135
Hyperstereology 189-200
Hypotenuse 9

Image analysis
  automated
    artifacts 127, 128
    capabilities 129
    comparison of counts with
      manual 129
    equipment 127
    feature analysis 128, 129
    on-line computer and 129
    preparation of specimen 127
    problems 127, 128
    spurious intercepts, sources 128

surface and volume density,
  determination 129, 130
  volumetric density estimates 128
semi-automated
  graphic digitizers 130
  graphic tablets 131
  on-line computer 131
  tablet-computer system,
    applications 131, 132
Index of folding or wrinkling (see
  folding)
Infinite extension of object 120
Interalveolar septal component 156
Intercept 22, 23
Intercept method for volume ratio
  determination 31
International Society for Stereology 16
Intersection 23
  counts
    direction of anisotropy 116
    glomerular filtration area,
      determination 39
    hepatic sinusoid surface area 39
    orientation axis 116
  unit area, number per 43
  unit length, number per 24
Isodiametric, renal glomerulus 8
Isotropic organs
  stereological methods 25
  surface area 37

Kepler's law 192
Kidney
  quantitative histology, method 119
  sampling 125
Lamina
  cumulative curve of width 83
  definition 83
  thickness 83
Length
  absolute 44
  concept 18
  fiber segment 45, 46
  internal 24
  internodal 24
  lineal elements 24, 41, 42

  fraction 24
  per volume, derivation of formula 42
  symbols 24
  total, gyri in cerebral cortex 44
  volume, determinations 124
Leukocytes, polymorphonuclear 47
Light microscopy, importance of thin
  sections 123
Linear structure, total length 41, 42
Liver cell nuclei, size estimation 122
Lung
  airspace chords 130
  airway identification technique 154
  alveolocapillary membrane, arithmetic
    mean thickness 158, 162
  alveolar surface area 157
  fixation and processing, correction
    factor, shrinking and swelling 152,
    156
  Fortran IV computer programs 162,
    Appendix II
  interalveolar septa
    differentiating components 158
    surface area 162
  intracellular constituents 158
  mammalian, stereological approach
    162
  maximum capacity of structure 152
  Parenchymal components
    light micrograph 161
    rules of recognition 157
  radiographic estimates of volume 152
  structural organization, mammalian
    lung 150, 151
  total lung capacity 152
  volumetric density 154, 156, 160, 162
  volume estimates 160
  volume-to-surface area relationship of
    bronchial vessel walls 154

Membranes 47
Microtubules, electron microscopy 124
Mitochondrion, basket shaped,
  sectioning 65
  shape of, determining 47
Morphometry, definition 17

Nebulae, barred or spiral 192
Nucleus
  columnar epithelia and fibroblasts 7
  liver cell, size estimation 122
  lobated and indented, shape 48
  truncated 122
Nucleotesimals 187

Opacity 121, 122, 189
Orientation
  analysis, coronary capillaries 116
  axis, intersection counting 116
  degree of, symbol 24
Ovarian follicle, volume ratios of successive developmental stages 35

Particles
  number
    symbol 24
    unit volume, number per 24
Pattern recognition program 129
Peripheral nerves, shape 6
Peroxisomes, diameters 122
Planimetry
  counting point hits 29
  Glagolev equation 29
  point count method 29
  serial, direct measurement 28
  slabs of equal thickness, procedure 28, 29
Pleomorphic mitochondrion 47
Podocyte
  renal, electron micrograph 2
  3-dimensional model 1, 2
Point counts
  methods 25
    bias, avoiding 26
    test patterns 26
  number per text area, symbol 24
  per unit volume, symbol 24
Polar caps, lost
  height, determination 64
  recovery of 68
  thickness and height 59, 64
Polyhedra
  cuboidal and columnar epithelium 9
  prisms 9

Profile
  counts
    capillary 116
    methods 25, 26
    number of, symbol 24
    number per unit text area, symbol 24
    tubules 41, 43, 44
  diameter, symbol 24
Prostate gland, volume ratio 29-31
Pyramidal cells, orientation 119

Radian, definition 13
Random variable 95
Reconstruction, tomographic images 147
Renal glomerulus, isodiametric 8
Ribosomes, electron microscopy 124

Saddle surfaces 15
Salykov's size distribution 78
Sampling considerations 124
Secant, definition and relationships 10-12
Section thickness 147
  definition 21
  determination 25
  Holmes effect and 124
Sectioning
  requirements for light and electron microscopy 60
  section thickness, estimating 61-64
Semitranslucent organs, images 147
Shapes
  branched sheets 54
  complicated, continuity vs discontinuity, solution, Leporello method 134, 135
  curvature, characteristics of shape 54, 55
  determination 41
  laminae, properties 54
  muralium 54
  parameters, index of folding, determination 52
Sheet, thickness in space, symbol 24
Size-distribution, methods 65

Sine 10
  curve 12
  definition and relationship 10, 11
Skeletal muscle fibers, prisms 9
Slab, series of equal thickness,
    measurement 29
  definition 21
  intersections and 22
  series of equal thickness,
    measurement 29
  thickness, symbol 24
Slice
  definition 21
  determination from 25
  thickness, symbol 24
  systems 17
  terminology 20
  volume, concept 18
Stereometry, definition 17
Stereo-photogrammetry 148
Surfaces
  absolute area of solid 40
  area per unit volume 37, 40
  cerebrocortical 40
  curved 20, 24
  curvature, mean, symbol 24
  density
    alveolar tissue, image analyzer 120
    basic programs 129, 130
    curvimeter 40
    glomerular basement membrane 39
    hepatic sinusoids 38
Swelling issue, during fixation 27, 28
Synaptic vesicles, electron microscopy 124

Taenia coli, elliptic cylindrical shape 6
Tangent, definition and relationships
  10, 11
Tetrakaidekahedra
  fat cells 9
  stratum spinosum of epidermis 9
Three-dimensional structures
  complicated shapes 134, 135
  identification from sections 134
Tomography 133, 135

Trace 22, 23
Triaxial ellipsoids 6
  (see also Ellipsoids)
Trigonometric functions
  changing signs 13
  (see also sine, cosine, tangent, cotangent, secant, cosecant, hypotenuse)
Trigonometry, stereological formulas 9

Unbiased estimator 95

Variance, estimating 100
Veins, elliptic cylindrical shape 6
Vision, panoramic vs stereoscopic 148
Volume
  concept, 18
  density determination
    automated methods 128
    basic programs 129, 130
    bronchial gland 34
    CAT estimates 147
    nonparenchymal and parenchymal
      tissue 156
    planimetry 28
    point hits 32
    Quantimet 720 image analyzer,
      basic programs 129, 130
  fraction
    core to shell 34
    tissue components 29-31
  ratio
    determination
      automated scanners 32
      importance in biologic specimen
        123
      intercept method, determination 31
      interference by Holmes effect 123
  terminology 20
  translucent finite thickness 121
Small-fold 60-62, 126
Solids 47
Sphere
  diameter 8
  histologic examples 75
  isodiametric 8
  opaque, overlapping of bounderies 121

Subject Index 305

    profiles 66
        classes, frequency 67, 68
        sizes, frequency 69, 70, 73
        Schwartz-Elias Partical Size
            Classifier 71
        sectioning 48
        size classes, determining 70, 72, 73
Spleen volume, CAT method 147
Stereogram
    definition 13
    description 3
    podocyte 2
    geometry of sectioning 3
Stereology
    aims 25
    applications 17, 189-200
    creating and defining points 17-19
    definitions 16, 17
    features in a test area, counting 18
    hematology, applications 197
    length, concept 18
    symbols
        DeHoff-Rhine-Underwood 23
        Michaels 23
    surface area per 37
    surface relationships of bronchial
        vessel walls 154

Weibel's system 23, 24, 44
Whispering galleries 5
Wrinkling
    (see folding)